CLARTS AND CALAMITIES

By the same author:

Don't Laugh Till He's Out of Sight
Funnyway'tmekalivin
The Magic Peasant

Clarts and Calamities

written and illustrated by
Henry Brewis

Farming Press Books
IPSWICH

First published 1988

British Library Cataloguing in Publication Data

Brewis, Henry
    Clarts and calamities.
    I. Title
    828'.91407

ISBN 0-85236-187-4

Published by:
Farming Press, 4 Friars Courtyard, 30—32 Princes Street, Ipswich IP1 1RJ, United Kingdom.

North American Distributor:
Diamond Farm Enterprises, Box 537, Alexandria Bay, NY13607

Typeset by MHL Typesetting Ltd, Coventry
Reproduced, printed and bound in Great Britain by
Hazell Watson & Viney Limited
Member of BPCC plc
Aylesbury, Bucks, England

Introduction

This diary bears little resemblance to those produced previously by Pepys, Crossman, Mole, or that Edwardian Lady.

It records nothing of great political or social significance (or if it does it's accidental). These are simply the daily scribblings of some umpteenth-generation peasant. There's no clue to his identity, only his circumstances,— and if the reader recognises a father, brother, husband, next-door neighbour, or even themselves ... fair enough.

This is a year in the life of a bloke who'll never drive a Porsche, seldom wear a tie, and doesn't commute to work, because he's there already.

He's been there since the first ewe lambed, the first cow was milked, the first field of wheat harvested ... and he still hasn't made any money (or so he says).

But he's still there.

Henry Brewis

'The farmer is the only man in our economy who buys everything he buys at retail, sells everything he sells at wholesale, and pays the freight both ways.'

John F Kennedy, Campaign Address at the National Ploughing Contest, Sioux Falls, South Dakota, September 22nd 1960.

January

A canny morning, cold and clear, frost rind on everything,
– very still. I might have enjoyed it if I felt m' normal self,
– but I didn't. It was Charlie's fault of course. Charlie and
Hilda farm next door, and we always end up at their place
to see the New Year in. He's one of those well-marinated old
gaffers who can soak up the pop, and it never has much effect.
Not me though. I generally know my limits, but last night
Charlie somehow dragged me way beyond them.

    There was little sympathy from Gladys, who was obviously
feeling fine, – she's skipped about all bleedin' day singing

1

'the hills are alive', — and the evil woman produced an enormous mixed grill for my dinner.

Slept most of the afternoon, and woke up in an even worse condition. Haven't had a cigarette today yet (I might stop altogether).

Fed the cattle after tea, — they seem to be blaring much louder than usual. Sweep appears to be in excellent health too, — but of course he wasn't at Charlie's last night.

Friday January 2nd

Haven't had a smoke since two o'clock yesterday morning, — but I feel a lot better today, and giving them up doesn't seem such a good idea now.

Fed the yowes after breakfast, — one short. Sweep finds her lying as stiff as a post in a rig bottom. To be honest I've been expecting her to go for a few days, — she hasn't come to the troughs for a week, and was looking like an anorexic goat. Now she's surrounded by a funeral party of crows, and the fox has already had a quick snack from the carcass.

The bitch, — she's had twenty quidsworth of wonder drugs since Christmas. So much for modern science!

Saturday January 3rd

I forgot to have the dead sheep removed, didn't I? — and of course the entire 7th cavalry (the Hunt) galloped right past the remains (not much left now). Naturally they frightened the rest of the flock into a heap in the far corner of the field, — except for the very lame ones, who struggled gamely to join their mates. I saw several riders laughing and pointing at the deceased.

After dinner, drew the curtains and watched Grandstand till they'd all gone. When I went to feed the bullocks, one of them had put his foot throught the ballcock, and there was a lake in the hemmel. Mending it was no fun, — water ice cold, all the bullocks licking m' backside and chewing on m' anorak while I was fixing the thing.

Sneaked the first fag of the new year, tasted lovely, — it's going to be difficult. Gladys says if I don't give them up, I'll have a heart attack. Charlie lost a collie dog once with a heart attack. I never saw *it* smokin'.

Nowt on television, never is on a Saturday. Fell asleep during a goal-less draw on Match of the Day.

Gladys woke me up at 11.30, — and so to bed.

Sunday January 4th

I always wake up a little apprehensive on a Sunday. All I want to do is read the papers, eat roast beef and Yorkshire puddin', — doze in the chair. When I went outside to feed the 'zoo', all seemed quiet.

But I should've known. The big hemmel was empty!

We eventually found the bullocks in the churchyard, terrifying the devout as they emerged from communion, and knocking down a few old gravestones. One mad charolais took a fancy to Mrs Simpson for some reason, and chased her back into the sanctuary of the church. The vicar reacted with very little Christian charity.

Had another cigarette when we brought them all home again, — but Gladys saw me, so I had to nip it.

Monday January 5th

Kids back to school. Willie (son and heir to the overdraft) will have to do better this year. If he doesn't stick in and do some real work, he could be condemned to a life in agriculture. Anyway now that he's not at home for dinner every day, perhaps I'll get the crust off the new loaf, and the small new potatoes.

Gladys watched a holiday programme on TV, and says she'll get some brochures tomorrow. This sounds serious.

Daughter Doreen goes back to the Poly secretarial course. She hopes to get a job in the bank some day, and I need all the influence I can get *there*, that's for sure, — but it'll be a while yet. Meanwhile I expect she'll get chatted up by a bunch of hairy, spotty left-wing students, all doing sociology.

Somebody called Melvin picks her up every morning in a yellow Citroen. He's got ginger hair and wears dark glasses, even in the house. He calls me 'Squire'.

Tuesday January 6th   *Epiphany*

This diary is becoming hard work already, — what the hell am I writing it for? Who on earth is going to read it? Gladys says keep going, 'cos it's for posterity, and might be a best seller when I'm dead. She says she'll sell up when I'm dead, go on a world cruise, meet a rich widower who doesn't smoke or drink or swear, and live happily ever after in Torquay, where it's warm. There's palm trees in Torquay, she says, and retired Generals and old ladies with purple hair, and string ensembles play Strauss waltzes while you have your tea.

The dog's been missing all day. I assumed it was off with the nymphomaniac Jack Russell from the village again, but it came back with Willie. Followed him to school, so he said. I don't suppose he discouraged it.

Wednesday January 7th

Mart day, — took two hours to get the hoggs off the turnips. They wouldn't come out through the open gate, — just ran around in rings while I swore and screamed at them, and Sweep got huffed and confused. Finally I shut the gate and loosened a rail in the fence twenty yards farther up. They went through *that* hole five at a time, — and set off up the road

(the wrong way) as if the hounds of hell were after them. By the time the last few were through, the first ones were a mile away in the village.

Eventually got a dozen right good'n's drawn off in the pens, and took them up to the grading.

The grader (known hereabouts as Ghadaffi, because the man's power crazy and y' can't argue with him) rejected five for being too fat. Charlie had a few rejected as well and we were both quite upset about it. Charlie threatened to assassinate the bloke by shoving his paint stick somewhere uncomfortable, but that attitude won't get him anywhere, — he'll just get more rejected next week, or (at best) four kilos knocked off. Anyway we couldn't understand what Ghadaffi's response was. He comes from Yorkshire.

Gladys has seventeen brochures specialising in everything

from Overland Safaris in the Hindu Kush to a British Rail weekend in Tow Law. There's one, I see, deals with holidays for the over-50's in Torquay!

Took the remaining lambs back to the turnips, but forgot I'd shut the gate. Of course they couldn't find the hole in the fence this time, and went straight past into the winter wheat.

It's snowing.

Thursday January 8th

Still snowing, — is this '47 all over again?

There's a hogg wrapped up in the electric fence, — the rest have trampled over the top of him, and found their way back to the wheat.

Forgot to turn the troughs in the back field yesterday, so naturally they're full of snow. While I was cleaning them out, the yowes knocked me down, and two of them managed to eat all the barley meal and nuts before I could get up again. These two will probably explode tomorrow.

Gladys fancies some Greek Island called Galapolos (or something like that, — they all sound the same t' me). I said we couldn't possibly afford it, reminded her of the five rejects, said it would be far too hot, — asked her who would look after the stock, etc, etc. But she didn't listen, — just went on turning the pages, and smiling a lot.

It's still snowing, we have about five inches already. If the wind blows we could be blocked in by morning.

Friday January 9th

We are, and the council ploughs won't come up our road until there's a thaw, — it could be Easter before *they* arrive!

Took the tractor to the pub. There were eleven tractors in the car park. Predictably Cecil Potts brought his new big Ford (he would, wouldn't he?). Charlie came on his Japanese trike.

He was perished, blue, — couldn't throw a dart for over an hour.

Doreen is in a desperate condition because the YFC dance has been cancelled, and anyway Melvin's old car isn't much use if there's a dew on the lawn, never mind a blizzard. Gladys says Doreen and the love-lorn Melvin have blathered all night on the telephone. I'm considering installing a coin-operated kiosk in the passage.

Sweep and the old tom cat are lying together in the kitchen, best of pals. They know if they create a disturbance they'll both be kicked outside, and it's a moderate night.

Saturday January 10th

There's a tax demand! Where did *that* come fron? I thought you had to make something called a 'profit' before you paid tax! And there's bills, hundreds of 'em! Right through Christmas you can't get in touch with anybody, — offices closed, everybody out celebratin', — only those bleedin' answerphones left in charge. Then as soon as the world starts up again, the first thing they do is send out a load of statements for all the stuff you bought last year, and had forgotten about!

We'll have to sell some more lambs. Only thirty-three left. Well thirty-one really, — two of them are collector's items, i.e. they haven't grown since birth, should've died long ago, lean as paper bags. But those little crits go on for ever, — just an embarrassing nuisance.

There's some barley going out next week, and that should please Mr Montague at the bank.

Sunday January 11th

Hogg dead on the turnips, a good fit one of course, never one of those rubbishy valueless creatures. I quickly decide to send the rest to the mart on Wednesday, before any more expire.

And the yowes are eating hay at a hell of a rate, — we could run out by March. Only once had to buy hay, and I remember it felt like burning fivers.

Gladys has a cold, or so she says. Y' know what women are like, — a bit neurotic, — always something wrong with them (it's the same with yowes isn't it?).

Willie sledging with his mates till after dark, — they include little Tracey Williamson, who gets a lot of attention. I fear Willie may have discovered women. Did little fourteen year old girls look like that when I was a kid? Maybe, — but they're certainly packaged differently now!

A lot of snow has blown into the hemmel, and it's a proper mucky mess in there. The bullocks look fed up, and are blarin' continuously. I think they're trying to tell me they need some bedding.

Monday January 12th

The council snow plough came past the road end mid afternoon, and it took Willie and me (Willie's school is closed because of the weather) two hours to dig out the snow they'd pushed into our gateway. By then the wind had got up and threatened to block the road again.

Gladys spends the day in bed. This is the second day she's had in bed since the war. I hope it's not becoming a habit!

Willie and I had to get our own breakfast and dinner. No bother really, but Willie complained the bacon tended to shatter when approached by a fork (well, I like it crispy) and the fried eggs looked like small pizzas. At dinner time we had some cold beef left over from yesterday, with potatoes and tinned peas. Perhaps the spuds were a little underdone, and undoubtedly the peas would've been better heated up. Willie said it was like eating yow pellets. He's spoilt, that's his trouble. Doreen cooked the supper, and I sneaked a fag when everybody was away to bed.

Tuesday January 13th

Very cold bright day. The countryside looks beautiful, but
the house looks as if a crowd of hippies have camped in it
for a year. The quicker Gladys recovers, the better. Maybe
I should be nicer to her, — or promise to look at some of those
brochures, — anything to get her up.

Bedded the bullocks with good barley straw. They went
crackers for a while, then ate most of it, galloped about for
half an hour playing games, — then they all lay down for a
snooze. They looked a lot happier, eyes closed, cud chewin'.
I wonder what they think about ... grass? heifers? high
protein nuts? ... holidays?

Wednesday January 14th

Mart day again. Brought the hoggs in with Willie and Sweep and a struggle. Most of the snow has blown off the fields, but the gateways are full. However they followed the tractor wheel marks all the way to the sheep pens, — me in front, boy and dog behind.

Sweep behaved very badly, and we had problems loading. Had trouble at the mart as well. To my utter amazement Ghadaffi graded the two rubbishy crits, and rejected all the rest! I'm coming to the conclusion I know nowt about sheep any more. What m' auld man would consider a decent finished lamb is too fat now. Those silly 'townie' women only want skin 'n' bone (if they buy meat at all).

Anyway the crits came to nearly forty quid a-piece, and a wholesale butcher from somewhere south of Darlington said he could do with a hundred like that every week! I just sold the others as stores, and they were bought by Cecil Potts. What on earth he's going to do with them is anybody's guess, — starve them I suppose. I'll keep a look out for them coming back to the grading, — but I suspect he'll sell them somewhere else, in case they lose money.

That clever bugger *never* loses money ... know what I mean?

Thursday January 15th

Gladys is much better, thank God. She seemed a bit disappointed at the state of the kitchen, but it wasn't that bad. Sweep was watching television on the sofa in the sitting room, and that upset her as well.

Willie is away playing with Tracey Williamson, — he seems to prefer her to driving the tractor now! It's funny how your tastes change as you grow older, isn't it? I can remember as a youth I hated cabbage, — love it now. Willie hasn't got to that stage yet.

Friday January 16th

Hard frost, — trough in the hemmel frozen, — cattle blaring for a drink, having smashed the ice and drained the trough. A kettleful of hot water does the trick. Trouble was I forgot about the kettle, and left it beside the trough when I went to put hay into the hecks. The cattle, desperate for a drink, crowded round and trampled the kettle to death. Gladys was not amused.

The roads are clear again, — Willie should get back to school on Monday.

Gladys went shopping. She reckoned that while she was in bed we'd eaten everything out of the freezer, and a hundredweight of chocolate digestives.

To the pub after supper. Three old-timers were discussing the worst winter they'd ever known. Their stories of mountainous drifts got higher and higher, as the beer went down and down. 1963 came into the reckoning, and 1947 of course, and another moderate year in the early twenties. They were tripping over telegraph wires, barely surviving on nowt but turnip soup, — livestock buried out of sight, — no light, no school, no food, no phone. No use interrupting either, — no sooner had one finished his tale of woe, than another came up with a more outrageous story to beat it.

From desperate winters, they progressed to other extremes, — plague, tempest, gales, droughts and finally money (or at least the lack of it). Somebody claimed he'd once sold fifty lambs for 50p a head (or ten bob in real money). The next fella recalled taking sheep to some obscure market on a bad winter's day, and finding nobody else had turned up. — just him and the auctioneer. The third bloke immediately remembered the time he'd *walked* thirty hoggs to Carlisle, where they stood in the main street for a week. Nobody took any notice of them, he said.

'So what did y' do?' asked some gullible idiot.

'Well we just left the buggers there, and came home ... ' said the lyin' auld divil, — without so much as a smile.

Saturday January 17th

Boadicea (Gladys's mother) comes and threatens to stay for 'a few days'.

We have a drunk bullock, and a dead yow.

There is no reason to suppose that these three events are in any way connected.

The bullock was one of several who got out of the hemmel overnight. Most of them went straight to the hayshed and pulled a few bales out, but this one found a bag of barley meal, and ate the lot. Robbie the vet came mid-morning and treated the intoxicated beast, − but he's still very wobbly (the bullock I mean, not the vet).

The yow? − well she very cleverly found a suitable hole in which to lie upside down, was no doubt delighted to discover she couldn't get up again, and duly expired, − her life's ambition realised I suppose, the sod!

Three fags today.

Sunday January 18th

A few flakes of snow this morning. Suggested mother-in-law should head for home before we're blocked in, but the snow had stopped by dinner time, and it was difficult to generate much panic when the sun came out. In fact Mr McCaskill is forecasting a thaw next week. All that means is more clarts.

Watched the farming programme. It was dead boring. All we got was a discussion between four 'teenage' bureaucrats from Brussels, all wind and waffle, rabbiting on about ESAs, clawbacks, restitutions and co-responsibility levies. I didn't understand a word of it. What I want to know is, how much will we get for the barley this year?

Which reminds me, — there's a load going out tomorrow.

Monday January 19th

Got up early to be 'organised', expecting first wagon at eight o'clock. Of course he came at twelve noon, just as I was sitting down to m' dinner.

I'm not very good with the grain bucket, but I had him loaded in less than an hour (which is very good for me), and was just sweeping up the mess, when another wagon came in.

The driver sat on his fat tail, eating his bait and smoking, while I shunted back and forward (ravenous) spilled a fair bit, and bumped his wagon twice, but got him away in forty-five minutes without serious damage. (This is a record.)

The worry now is this, will they find anything wrong with the grain? A few beetles in it perhaps? A dead rat? Too much straw? Dust, gravel, weeds, wheat, — an old sandwich maybe? The trouble is the price has dropped a fiver since we sold it, and you can bet your life they'll try to wriggle out of the contract if they can. If the price had gone up, I might've been disappointed, — but there'd be little worry about any rejections.

However no nasty phone calls by supper time. Relief, but slept badly. I was 'loading corn' all night, and Gladys was snoring several decibels higher than usual. She denies this of course, — but how the hell does she know? — she's unconscious. It was like a herd of newly-spaned suckler cows in full cry.

Tuesday January 20th

The alcoholic bullock is much better. That's the good news. The not-so-good news is that Boadicea (Gladys's mother) won't leave until the end of the week. I hate people staying, — I'm obliged to be polite, eat properly, no rude noises after supper. The bathroom's permanently engaged, — I can't be miserable whenever I feel like it. The whole thing is a violation of my air space!

The wagon came for the last load of barley when I was miles away feeding the yowes. By the time I got back to the shed the driver had done nowt, — just sat in his cab drinking tea, and listening to some jibbering disc jockey on Radio One. He was only about twenty, but still as fat as a pampered tup. (Come to think of it, the disc jockey was probably even younger and fatter, and a lot richer.)

Wednesday January 21st

There are reports of rustling in the area. Charlie from next door came to borrow some diesel, had a cup of coffee, and told us that his cousin Malcolm had found his entire farm in disarray yesterday morning, — livestock all over the place in an agitated state, and half a dozen hoggs missing.

The police suspect there's a gang of 'townie' rustlers at work.

Charlie and I are not convinced about this. I mean can you

really see a couple of 'wasters' from Mafeking Gardens being successful at this sort of thing? There they are, a garage full of hot videos already, deciding that a spot of sheep stealing, Clint Eastwood style, would be dead easy, and a nice way to supplement the dole.

So, with a bag of ewe and lamb nuts, and the inevitable lurcher, our heroes drive their old Transit out into the night, in search of instant mutton. The simple peasants are all at kip, the countryside is still. The subtle plan is to tempt some unsuspecting sheep to the troughs, and grab a few while they're feeding, — it shouldn't be too difficult.

However, little do they realise that their intended victims are the well-known ravenous mule yow variety, and before they realise it, our amateur rustlers are completely over-whelmed in the rush, buried in the clarts, the food consumed in a trice, and the bewildered lurcher has fled. One bloke maybe makes a despairing grab at some lean lame straggler, but all he's left with is a handful of wool as she splutters past.

Alright, so they feel they've got to steal something before they go home, — how about some cattle?

Well, just up the lane they come across a slumbering herd of suckler cows and calves, lying munching in the moonlight. Silently they climb the fence and creep in among the cattle. The cattle don't seem to notice, nothing moves.

'Here's a little black one!' whispers one of the rustlers, 'let's grab it and get out of here.'

It takes this dozy calf about two seconds to wake up and realise that some fool has hold of his tail, at which point the calf calls for his cross-Galloway mother. Mother arrives at the speed of light, bellowing like a Banshee, and accompanied by all the other concerned mothers as well!

The 'cowboys' flee (who wouldn't?) and the Galloway destroys the abandoned Transit, as the petrified lurcher tries desperately to hide in the glove compartment.

Charlie and I conclude that rustling can only be successful if performed by professionals, — like ourselves.

Thursday January 22nd

Still no barley rejections. We should be OK by now, so I ring Northern Grain for the weights, and add a rather pathetic plea for early payment. Consider sobbing into the phone, but the man hangs up before I can get started.

I may buy the Nitram elsewhere.

Some farmers will never admit to being hard up, even with a rampant overdraft and, unable to purchase a Mars bar except on credit, they'd still have you believe the livin' was easy. Others would plead poverty, even if their granny had left them a gold mine. Personally I always claim to be destitute, — that way Gladys doesn't expect too much.

She's still twittering on about the holiday though.

Friday January 23rd

A Blizzard. Flakes the size of fag packets coming in from the barren wastes of County Durham, — they're very wet however, and not really lying, — just a very ordinary day. The yowes are standing with their bums to the weather looking miserable. Even the bullocks in the hemmel have gone into a huddle at the far end, out of the wind and wet.

Got everything fed, and went to Charlie's for coffee. He's not in, — he's gone to Arthur Thompson's for coffee, so I have to listen to Hilda telling me how their George has seven 'O' levels, and his sister Sheila has a brilliant job in a solicitor's office. How is it that everybody else's kids are geniuses?

Florrie, the old bat who comes in twice a week (Mondays and Fridays) to help Gladys, and who has six children of her own, — says it's all to do with the genitals of the parents. She could be right of course, but I think she probably means 'genetics'. Anyway it's my fault!

Boadicea goes home. Turned out to wave her good bye, and had a celebratory fag.

Saturday January 24th

The big tractor has a flat tyre, the other tractor has a flat battery, — so I'm forced to carry feed to all the sheep on foot.

After breakfast set off with wheel and battery to Johnson's garage, and come face to face with 7th Cavalry. They're meeting outside the pub, it's chaotic. The Colonel is screaming at everything to 'get orf the road', little pot-bellied ponies are biting each other, ladies with massive thighs and sprayed-on jodphurs are larfing loudly from under hard hats. The Huntsman is bellowing at wayward hounds as they sniff about, peeing on everying. And all the time the horses are dancing sideways like four-legged Fred Astaires. There's a real risk of having the side of the pick-up kicked in as I crawl through among them, but I crouch low in the seat, and keep going in second gear.

When I got past I just couldn't resist a good pee-eep on the horn, and looked back to see Colonel Nicholas on his demented gelding shoot off into the car park, 0—60 in about two seconds. Apparently the same horse can go from 60 to 0 in about two seconds as well. Old Knickers can't quite manage that of course, — and tends to carry on alone. One beast with a red ribbon on his tail kicked another horse in the belly, and two more jumped into Mrs Simpson's garden from a standing start, — but by then I was through and away.

Lit up a secret cigarette from a packet I keep under the seat, — great!

Sunday January 25th

We have a lamb!

It has just appeared, as if from a magician's hat, — though of course I'm fully aware that it came from somewhere else. It is the smallest, runtiest thing I've ever seen. It is alive, but completely alone. Nothing, nobody, is willing to accept respon-

sibility. Some yowes give him an inquisitive inspection (as if they weren't sure what it was), but none declare a proper maternal interest, and there are no tell-tale signs with which to identify the sneaky mother either. One of them must be the culprit, and I tell them straight, in no uncertain manner, that if she doesn't come forward and own up immediately, I'll beat her to death when I find her. This little speech (expletives deleted) given with great feeling and conviction, has no effect. There are no volunteers.

I had just started to threaten the entire flock with slow starvation when there was an unmistakable snigger behind me. Charlie was leaning on the gate listening to it all. Very embarrassing, — he thinks I've flipped (again).

Monday January 26th

The runty lamb still lives, though with little enthusiasm. He spent all yesterday standing on a cardboard box by the kitchen fire, bleating miserably. Presumably his complaint is that he expected a somewhat different lifestyle, and is confused to find his mother looks like Gladys. Be that as it may, she managed to get some colostrum into him, and although the pathetic creature just stands there shivering, all four feet on the same spot, back up, dribbling continuously into his soggy cardboard home, — he *might* make it.

His proper mother remains anonymous, but I swear I saw a four-crop mule grinning this morning.

Tuesday January 27th

One yow missing. Naturally I assumed she'd died somewhere, but no, — there she was in the far corner, — with a pair, — both on their feet, bright as buttons, full as eggs!

Delighted though I am to find this happy family group, I'm beginning to wonder where it will end, — we're not supposed to start lambing until March 22nd.

Gladys solves the problem. She reminds me that Charlie's tup got through the fence one day last October. We chased him back within five minutes, — but obviously he didn't indulge in much fore-play.

Wednesday January 28th

Gladys crashed the car today.

Actually it's a miracle it hasn't happened before now, 'cos she can't really see where she's going. It's not that she's blind, no, — the problem is she's quite a short woman, and it's difficult for her to reach the clutch or the brake, and look out of the windscreen at the same time. She tends to slide off the seat as she stretches down. Even with a cushion under her bottom, she still has to peer up through the spokes on the steering wheel at an angle. Consequently although she can easily see a low-flying phantom jet swooping in to attack, any poor sod on a bike quietly peddling down the road is always at some risk. For years she's meandered into the village to do some shopping, or take the kids to school, smiling sweetly through the wheel as she strains to see over the dashboard. She is always totally unaware of anything to the right or left or behind, and figures those in front have an obligation to get out of the way.

Gear changing obviously demands a considerable effort, so she confines this exercise to top and reverse. All standing starts therefore demand maximum revs, full choke, and the clutch half-way out. This technique is also used at traffic lights and pedestrian crossings. People tend to cross very quickly when Gladys is poised there, engine roaring, the car easing forward, apparently without a driver. Local ladies grab their children and drag them across, others wait until she's gone. When she goes, she goes like the wind.

Only the good lord and St Christopher know how she hasn't caused a swathe of death and destruction throughout the neighbourhood, but she's never had even a minor bump (until today). While everyone else ricochets from one insurance claim to another, little Gladys has just stuttered around the parish, grinning through the steering wheel, totally unmarked.

But today she hit Charlie's Japanese trike as he was going to feed some sheep. She would never see him of course, but luckily Charlie saw her, and baled out into the ditch, followed soon after by the oriental trike.

Nothing serious, — the trike is bent a bit, Charlie is complaining about his bad leg, our car has a broken light.

Gladys is sitting in the kitchen drinking brandy, and insisting Charlie should be breathalysed and have his eyes tested.

Thursday January 29th

The yow and the pair are doing fine.

The orphan runt is looking a little less peculiar now. The humpy back is coming down, his legs are gradually moving out to each corner, and he doesn't bleat all the time. He's a good sucker, in fact he sucks everything, — chair legs, trouser legs, Sweep, the cat, — anything. He would undoubtedly empty a forty-gallon drum of milk if given the chance. All this intake goes straight to his elastic belly, and after each feed he looks like a pregnant rat.

Later this intake is released of course, so Gladys has moved him into a corner of the stable. Sweep, who also kips in there, looks huffed. Perhaps, like me, he doesn't appreciate uninvited guests who fracture his privacy.

Friday January 30th

I half-expect another birth every morning, — but I think that's probably the finish of the early lambin', — after all twice in five minutes is pretty good goin', especially if you've just fought your way through a thorn hedge.

Gladys didn't want to drive herself to the shopping today, she's still shaky after the bump, but I told her that Nigel Mansell would get straight back in again, before he had time to consider what might've been. She says it's much easier for Nigel Mansell, because in formula one racing it is very unusual to find someone else coming the other way on a trike.

Anyway she set off eventually, top gear, engine roaring, peering nervously out through the wheel (not grinning this time). Sweep stayed close to heel until she was out of the yard.

Saw Jimmy Fletcher from Windyhaugh in the Swan. He and three other blokes (a solicitor, an accountant and a bank manager) are all off to Spain to play golf tomorrow. I think Jimmy's out of his depth with wealthy people like that. On the other hand he's obviously been saving up for a long time, — hasn't bought a round of drinks since Newcastle won the Cup.

Saturday January 31st

Gladys reminds me that it's Willie's birthday today. She seems surprised when I ask how old he is, and refuses to tell me. I think he's fifteen, no, he can't be, ... can he? ... let's see ... he *is* y' know, ... he's fifteen already!

Ye gods, I think my memory's going. Actually I've suspected for some time that I might be suffering from some form of premature senility. I blame it on a lifetime spent in the company of sheep.

Anyway, I couldn't forget Willie, — so I nipped down to the shop and bought him a ball-point pen and a pocket calculator. He needs all the help he can get.

When I got back Gladys said that was typically mean, and sent me out again for a new pair of wellies, and a box of chocolates. She just spoils the little divil.

When I was his age ... .

February

Sunday February 1st

It's raining, — it woke me up, belting on the window. Most
of the snow has gone, — lagoons everywhere. The yowes are
marooned on a rig top, the troughs floating about. The tractor makes a proper mess wherever it goes. Sweep is in one
of those awkward moods when he doesn't fancy getting his
feet wet, — tip-toeing about like a bloody poodle.

Watched the farming programme and Mr McCaskill's
forecast. The fool's talking about spring already.

Charlie and Hilda came for supper. The kids played pop
music in the kitchen at a million decibels. I swear the house
was vibrating. Seems they can't live without that noise these
days. Even when Willie's doing his homework, the 'music'
has to be roaring. How can he think? Obviously Sweep can't
stand the racket either. We can hear him howling in the stable.

Monday February 2nd

NFU meeting in the village hall, — some bigwig from head-
quarters is speaking to we peasants. Our branch secretary has
been trying to persuade everybody to go, because there's a
crisis in agriculture. (So what's new?)

I went along, agin m' better judgement. I knew I'd be
bored, and the hall is always freezing.

The Chairman spent about half an hour saying how
delighted we were that Mr Bigwig had come all this way to

speak to us, and what a very distinguished bloke he was, and how privileged we were, and how eager everybody was to hear what he had to say. And then for the next two hours this bleedin' distinguished guest droned on and on about white papers, green pounds, and a black future.

Turns out he had a farm in Lincolnshire, about a thousand acres (which he saw on alternate Sundays), — growing and breeding all kinds of exotic things, like carrots and cauliflowers, as well as a million pigs and two million hens. I figured we didn't have a lot in common, — he had a suit on for a start!

I must've nodded off, because when Charlie nudged me the man was inviting questions. Cecil asked one (he always does), — but everybody else was sliding out to the pub.

We just made it for last orders.

Tuesday February 3rd

One of the very lame yowes has been trundling to the troughs on her knees for weeks now. She's quite easily caught, so I've been dressing her feet, and spraying that smelly purple stuff on, — but she's not much better. Now, it seems, her knees have gone.

I waited and waited for the arthritic auld bitch to get to the feed, but I could see it would take all day, so I fed the rest, and kept a bit back for her. You've guessed it, — no sooner had I put her wee ration down on the ground in front of her nose, than the others left the troughs and galloped across. Both the cripple and I were knocked over and submerged.

Perhaps I'm not giving them plenty.

Wednesday February 4th

Michael McMurdie, the autioneer, rang up. Do you ever get the feelin' that England's been invaded (again) by the Scots or the Irish? Percy Robson (who admittedly has some rather right-wing views) reckons every trade union official is a Glaswegian communist, and should be carted off to Siberia, — and all the comedians, weather forecasters, and political correspondents are IRA spies, and they should be tied up in plastic bags and dumped in Galway Bay!

Anyway McMurdie says he's short of fat cattle this week, and could I send a few to the mart? — 'they'll be a flyer', he says.

I should've known. He ended up with about thirty more cattle than usual, and three fewer buyers. The three who *did* show up stood telling each other jokes all day, and took turns at the bidding.

I sold two bullocks and brought two home again. We'll call it a draw.

Willie off to a YFC meeting at night. For years he hasn't

been interested, — said they were all a bunch of 'posers' (whatever that means) and boring (that's the *in* word for anything he disagrees with). Now he's changed his mind it seems.

We hear Tracey has joined as well.

Thurday February 5th

Telephone out of order. Willie can't phone Tracey, Doreen can't get in touch with Melvin, Gladys can't gossip to the massed blatherers of the WI. It's a crisis, — but it could save me a small fortune.

Friday February 6th

All the pigeons in England are grazing the oilseed rape, — stuffing themselves. They're like a flock of grey locusts, — you can almost hear them munching.

Got the old 12-bore out, and fired a few volleys at the dense cloud, but nothin' fell down. Put the banger next to the wood at the end of the field (it makes a bigger bang when it's under the trees) and set the clock for thirty-second intervals. This means that the machine goes off like a Gatling gun. If nowt else, it might give the pigeons a headache. In fact the only hope is to persuade the sods to go somewhere more peaceful. Next door will do nicely.

The British Telecom man came. I had a long talk with him, but he said he was obliged to mend it.

Willie went to a disco at the rugby club, dressed like a Chicago gangster. Doreen and Melvin were going there too, but Doreen said she couldn't be responsible for bringing Willie home by twelve. So I had to pick up the little waster.

As I drove up towards the clubhouse, I saw him in the headlights leaning over the fence being sick. My god, — he's a teenage alcoholic as well.

Saturday February 7th

The Hunt are on the rampage again. I think they met at Arthur Thompson's this morning, but by dinner time they'd chased some lame dog fox into the trees next to our rape field. There they were, quietly drawing the wood, 'other ranks' taking the opportunity for a swift sip from their flasks, patiently waiting for those dumb dogs to pick up a scent, . . . when the Gatling gun went berserk.

Alright, maybe I should've switched it off for the day, — but I have to admit the chaos was a joy to behold. That bird scarer can shift more than pigeons.

Willie is off his food. Gladys thinks he's coming down with some bug or other, — 'there's a lot of it about', she says. She was just on the point of giving him a dose of syrup of figs (her cure for everything from anthrax to a broken leg), when I told her he was 'emptied' last night.

Sunday February 8th

Gladys went to church, — perhaps to pray for Willie.

I fed the livestock. It was one of the bright, crisp, clean winter mornings, when the countryside and everything in it looks sharp and healthy. There's a crunchy rind on the ground, a film of ice on the puddles, a filigree of frost on the branches, and breath hangs in the air. A joy to be alive.

Apparently the yow with the arthritic knees didn't agree with this somewhat romantic view . . . and has expired.

Monday February 9th

It is half-term, so Willie has a full week to recover. His mother suggested the 'poor thing' had a day in bed. But I dragged the little wretch out when she went for the papers, and made him bed the bullocks and feed Womble the wonder lamb (that's the motherless creature born on the turnips last month).

The early arrivals are doing pretty well, though Womble is now virtually *all* belly. Chances are we'll feed him at great expense and considerable inconvenience until the grass comes, then he'll explode in a cloud of milk subsitute and protein pellets. That's if the postie doesn't run him over first.

McCaskill is forecasting snow again.

Tuesday February 10th

Cecil Potts, our local agricultural genius, has starting lambing.

He's one of those blokes who always has a 200 per cent crop and no bother.

If he *has* a dead yow (and he won't admit to that) presumably he buries it himself in the dark. He generally buys and sells everything privately, so no one can argue when he claims his purchases were a stroke of entrepreneurial brilliance at rock-bottom prices, or his sales cleverly timed to coincide with a roaring trade. He considers marts to be an anachronism (whatever that is) — but what he really means is he doesn't dare go there and bid like everybody else, in case he makes a fool of himself in public.

If he's the sort of bloke the Ministry rings up to get some guide on crop yields and the like, no wonder they're worried about surpluses. But then, as Charlie says, 'we know he doesn't fly every time he flaps his wings ... !'

It's not snowing yet.

Wednesday February 11th

Took six bullocks to the mart. Canny trade, — by which I really mean I got about ten quid a head more than I expected (though I'm not all *that* sure what I expected).

Charlie had a reject, — so I felt pretty chirpy.

Still no snow. I wonder if McCaskill knows what he's talking about? Gordon, who helps at the mart, and is the meteorological guru in these parts (he's an authority on the behaviour of slugs, seaweed, hedgehogs etc.) says it'll snow tonight for sure, because he saw a lot of snails creeping across the main road yesterday. That's a certain sign, he reckons.

Thursday February 12th

A soft balmy morning, a slight drizzle. So much for Gordon's snails, and the McCaskill computer.

We have a massive bill for coal and Aga fuel, which inspires me to cut logs all afternoon with the chainsaw. I also manage to cut the toe end off m' left wellie. I blame Arthur Scargill.

Friday February 13th

Charlie came down for a lamb to set-on. He's had a yow who hanged a ruddy great single last night, and he offered to *buy* Womble. This was an astonishing gesture coming from Charlie, who (like me) comes from an era when spare lambs were given freely to needy neighbours. But now it's a mercenary world, and everything has its price.

I charged him a bottle of whisky, and was pleased to see the back of the little runt (the lamb I mean, not Charlie). Whether I'll ever see the bottle of whisky is another matter altogether.

Saturday February 14th   St. Valentine's Day

Gladys in silly mood, presents me with a card and a sloppy kiss at breakfast time, which is most disconcerting. Anything more demonstrative than a cough at this time of day is considered excessive in this house. To be given a Valentine (all hearts and roses) *and* a kiss, albeit on the top of m' head, was extremely worrying. Was she feeling guilty about something? Is she after a new frock? Or a three-piece suite? Has she bought these things already?

Anyway I said nowt, downed a cup of tea and went to feed the bullocks. As I went out the door she shouted, 'Y' miserable auld git!' . . . and I felt a lot better after that.

The postie came with a card for Doreen (from that drip Melvin I expect) and four for Willie! That boy is going to be a problem.

Sunday February 15th

Willie is playing for the school against the Old Boys XV today, a special annual match. He's pleased to be picked. Perhaps

sport (and a few cold showers) will help to divert his sordid little mind.

I went along to watch the game, and was dismayed to find that the only other spectators were four nubile maidens dancing merrily up and down the touchline, dressed in mini skirts, black wooly tights, small sweaters, and large bobble hats, all screaming 'get stuck in Willie', 'kick 'im Willie', and other demure suggestions.

To be fair, Willie did his best to oblige them, and was duly sent off early in the second half, — to rapturous applause from his fans.

Monday February 16th

Mr Montague from the bank phoned and asked me to come in for a chat, — 'nothin' serious', he says, — 'just a little discussion on what facilities you may require over the coming twelve months.' 'Facilities' is a banking euphemism for money (they never discuss 'money').

I told him I was extremely busy for the next week or two, but would phone back and arrange an appointment on a mutually acceptable date. 'A mutually acceptable date' is *my* euphemism for the day I get the barley cheque paid in, and a few more bullocks sold.

It's snowing tonight. So Mr McCaskill and Gordon were right after all, even if it is a bit later than they, and the snails, had anticipated.

Tuesday February 17th

Woke up to a level three inches of snow, — no wind, but very cold. Let's face it, this is a lot better than rain and clarts.

Nevertheless all the soft townies south of Darlington are panic-stricken and bewildered by this amazing development (i.e. snow in February). According to the news, there are trains marooned on frozen points, yuppies sliding about on the M25, jack-knifed lorries blocking the M6. Teachers are sending the kids home from school after the coffee break, offices and factories close, buses cower in the depot, newspapers never get beyond Watford, — and the Government get the blame.

Meanwhile in the real (rural) world, simple peasants are de-frosting water troughs, feeding the yowes, bedding the cattle, mending the fence, shovelling muck, . . . or whatever. It's just another Tuesday, and the winter ain't finished yet.

Wednesday February 18th

Snow showers continue. Give the yowes a bit extra food, which they consume in a flash. Sweep seems to enjoy the snow, sticks his nose in it and snorts happily. He's reluctant to 'sit' though, and just hovers with his back legs bent, and his bum an inch above the snow. He has a pained expression on his face until he's released from this position, and free to run again.

Took Doreen and Willie by tractor to catch the bus in the village. Willie protests that the school will be closed, but I know he's out of luck. Doreen insists that the Poly won't need her today ... but she goes too.

And four bullocks go to the mart. Only two buyers turned up, but thankfully not many cattle turned up either, and we got a fair trade.

Thursday February 19th

Sitting having a cup of coffee when the phone rings. Fortunately Gladys answers it, because it's Montague again. 'I'll see if I can find 'im' she says in her posh voice, and then covers the receiver with a cushion while she tells me who it is (she's well-trained in such procedures).

'I'm sorry Mr Montague, but he must be away on the tractor somewhere,' she says (I'm sitting like a statue) ... 'can I take a message? ... Right, Mr Montague, I'll tell 'im' (I feel a sneeze coming on) '... he should be back by dinner time, ... yes, it's a busy time of year ...' (I creep into the passage and explode '... no Mr Montague, it's just the dog barking ... thankyou very much then ... bye.'

Phone grain merchant immediately to hurry the grain cheque. Of course the boss is on holiday in the Bahamas, but his secretary promises it'll be in the post by tonight. (She sounds almost as convincing as Gladys.) I think she's Arthur Thompson's daughter. I'll have to have a word with *him*.

Friday February 20th

To the pub with Charlie and other peasants.

There's a darts match with the Plough which gets rather noisy. They have a McEnroe-type player who hurls abuse every time he hurls a wayward dart. There's talk of him being barred for the rest of the season, but this suggestion only seems to make him worse, until Elsie refuses to serve him. That cools him down.

Charlie, m'self, Percy and Arthur have an in-depth discussion on the current state of the world in general, and farming in particular. Percy, who has a very simplistic view of life, reckons everything is on the brink of disaster, and the only solution is the immediate reintroduction of hanging, — especially for terrorists, rapists, murderers and *solicitors*. (He's in the middle of a long legal battle after backing his John Deere into the postman's van.)

The rest of us, joining into the spirit of the thing, add rustlers, VAT and safety inspectors, government graders, and several reps, to the list of doomed men who must be sacrificed for the sake of civilisation, come the peasant's revolution.

Farming, we conclude, is a knacker job, and only congenital idiots would pursue such a career.

The congenital idiots, having by now consumed six pints a-piece, disband happily at 11.30 pm.

Saturday February 21st

A fair amount of snow has gone, the rig tops are bare, — and the 7th Cavalry is out again. (We forgot about them when we compiled our revolutionary death-list last night.) Fortunately they just trot through towards Charlie's place, smiling their 'mornings' as they go. They appear to be in good form, I can hear a lot of supers and spendids, and even a few *absolutely* supers and splendids. I'm reminded of the Light Brigade cantering carelessly towards the Russian guns. Unfortunately this brigade will return more or less intact, — but I may have a word with the Colonel and ask him to stay clear of my pregnant yowes from now on. There'll be enough problems without his happy band stirring them up.

Willie went to a YFC disco with Tracey, — apparently he is forgiven for being sick at the rugby club, and romance blooms again. Melvin and Doreen will be there too, so I ask daughter to keep an eye on wayward son. She says this could spoil her night.

I go to pick him up at midnight. He's not there, nobody has seen him (or Tracey) for over an hour. I'm sitting in the car park, quietly fuming, tired and cold, having a sneaky cigarette, — when I'm aware of movement in the next vehicle. A steamy window is lowered, and there's Willie with a can of lager in one hand, and a piece of Tracey in the other. Ye gods, he's only fifteen and already he's on a par with Rasputin!

Just get to sleep when Melvin brings Doreen home and wakes me up. Gladys is making hippopotamus noises. Being a peasant parent is not easy.

Sunday February 22nd

Gale-force winds. I don't like winds, every job is harder in a wind. Maybe sometimes it's useful for drying up clarts, but generally it just chews and tousles y', — blows the heads off the barley, blows straw all over the place, brings down trees which take half the fence with them. Windy days are generally angry, bad-tempered days.

Today it was the empty plastic bags that were on the move, and several had found their way into the hemmel. The bullocks were having a grand time chasing them as they flew about like demented kites.

You have to be fond of cattle, — unlike sheep they have a sense of humour, no matter what age they are. Perhaps you could argue that healthy lambs can enjoy themselves, playing chasey up and down the dyke back. However, as soon as they grow up they tend to become dull and boring, and set in their ways. But cattle are inquisitive, actually display a degree of enjoyment and they're capable of simple jokes. And what's more they generally have 'livability'. A sick beast *can* be cured. You've got a chance, he'll generally respond.

Sheep aren't half so enthusiastic about life, — perhaps they're deeply religious, and eager to be in a better pasture elsewhere. Or perhaps they're just bloody awkward!

Monday February 23rd

The yowes look quite well (by our standards anyway, —
everything's relative). It's the wind that does it of course, —
fluffs the fleece up and out, makes them appear fitter than
they really are. Good hoggs on the turnip break can look fantastic on a dry windy day, — full and yellow.

Had a good check on the feed supplies, — the hayshed is
getting low, must keep selling cattle. Plenty barley left I think.
And the cheque has arrived. I'll be almost 'flush' for a while,
all the corn payments in, and still a few bullocks to sell.

Then the miserable voice of common sense tells me that
I'll have it all spent in a week or two ... store cattle to buy,
seed, fertiliser, spray, concentrates, — not to mention the
usual phone bills, electricity, coal, repairs ... and a few other
items that I'll have forgotten about, bills that come and grab
you, just when you're feeling pleased with yourself.

Which reminds me, — Montague phoned again. 'No problem,' says I, 'how about Friday morning, shall we say eleven
o'clock, see you then, ... bye.'

Rushed to the bank and paid in the barley cheque.

Tuesday February 24th

Gladys is out tonight, — a WI trip to see the *Mikado*. Willie
thought it must be some ageing pop group. 'It's Gilbert and
Sullivan', says I. 'Were they before Lennon and McCartney?'
he asks!

I fear there is little hope for the ignorant youth. What on
earth do they teach them at school these days. His writing
is terrible, resembles the work of some aborigine cave artist.
Reading? ... I've never actually seen him reading anything
other than *TV Times*. Arithmetic? Well at least I know he can
count the yowes if they stand very still (and most of our yowes
can do that alright ...)

Wednesday February 25th

Six bullocks to the mart, all graded, canny trade.

Arthur Thompson had a right wild thing there. I don't think it was properly castrated, a real 'bully' head on 'im. Anyway it came roaring out of the wagon before the tailboard hit the ground, and galloped up and down the alleyways for most of the morning. By the time Gordon and his gang had him penned, he was 'guffed', but still looking in a proper bad fettle. Nobody fancied handling him, and he was graded from a distance (in fact he was lucky to be graded at all).

By the time it was his turn to be sold, he'd got a second wind and went crazy again. Seemed determined to destroy the weighbridge, and they had to give him an 'estimated' weight, 'cos the needle on the dial wouldn't stand still long enough. Then they opened the doors, and in he came like a Spanish Toro. The ring cleared like snow in a fresh. Gordon, who's nearly seventy, was over the rail with one leap. Two over-fat butchers, who normally just waddle about like performing seals, took off like Peter Pan and Wendy, and settled in the auctioneer's box.

Michael McMurdie had the beast knocked down to one of them after its first circuit, in spite of Arthur's protestations. But Arthur was in the back row of the gallery with the rest of us by then.

Some brave soul opened the gate and El Toro was out, still snorting and looking for something to thump. We all moved cautiously back down to the ringside, and the sale continued.

Thursday February 26th

A wet night, — snow disappearing, but still a bit left along the hedges. Gordon reckons it'll be there for a while yet, because it's waiting for some more. I assume he must have seen a couple of rabbits dancing an old-fashioned waltz, or something prophetic like that.

Be that as it may be, — the yowes are in a mess, troughs in a sea of clarts.

I spend the day converting the grainstore and the empty part of the hayshed into a maternity ward. It won't be long now.

Friday February 27th

The big day, — Montague awaits.

Got to the bank quarter of an hour early, which was a mistake, because he still had another 'victim' in his office, and I had to wait about like a hat stand, looking nervous and conspicuous. It was obvious why I was there. I read the exchange rate, 196 pesetas to the pound. I read how easy it was to get a loan if you were a destitute student. I read how to apply for a credit card, a cashpoint card, a home loan, a car loan, a bleedin' holiday loan! They were positively desperate to lend everybody some money. So what was I worried about?

Montague emerged at 11.05 preceded by a flushed Cecil Potts, who scurried out into the street like a scalded cat. So (thinks I) even the big lads have their problems, eh? — and maybe their problems are bigger.

Actually the interview went pretty well really. He called for two coffees and my dossier. We went through what he called assets and liabilities. He listed the yowes as assets (though that's a matter of opinion). And the bullocks? What were they worth, he wanted to know. (I gave him the price I got for the best 'n', — and didn't mention the little pot-bellied thing with ringworm.)

'Any grain left?'

'Oh yes,' says I, and gave him a tonnage that would've fed Ethiopia for a week.

'Hay?' 'Straw?' 'Machinery?' he asked, moving down the page, and I gave him a comprehensive list of all our antiquated gear, making it sound as interesting as possible. I included everything, even the old acrobat and the zigzag barrows currently blocking a hole in a hedge.

He offered a figure to cover the lot that left me breathless, which he apparently took to be a surprised reaction of gross *under*valuation, — 'cos he immediately added a bit more.

'Well,' says I, recovering my composure . . . 'that seems fair enough.'

We moved on to liabilities, and I 'forgot' to tell him that I hadn't paid for the load of Nitram, included as an asset earlier on. It seemed unnecessary to worry him with the water rates, and a few other 'outstanding' problems.

By the time I left at quarter to twelve he seemed happy enough with my story, and when he suggested the all-important upper overdraft limit, I very nearly tore his arm off.

Had a celebration drink with Charlie and company in the Swan, but said nowt about Cecil.

Saturday February 28th

Willie playing rugby again, so dropped in to watch the second half. No groupies this time, only Tracey, and she was chatting up the lad running the line.

Willie scored a try, one of those quick ferrety tries that scrum halves sometimes score, but he got clobbered by the whole pack with ten minutes to go, and had to be helped off, apparently suffering from concussion. When I went over to see if he was hurt (I must admit, I was a wee bit worried) — he had a glazed expression and was groaning quite a lot (perhaps a bit *too* much). Anyway I ran off to get a coat from the car to wrap him up, and when I came back the little bugger was beating the daylights out of the touch judge, — the one Tracey had been flirting with. The boy's impossible.

March

Sunday March 1st

A fairly lazy Sunday. Gladys resurrects the holiday issue, even switches Harry Secombe off, and insists on a proper debate (with her as chairman of course, — sorry, chairperson). There you are, see? — even I'm influenced by this woman's lib rubbish, — it's destroying decent marital relations. M'father would never have stood for a 'proper debate', and mother never complained. They never had a holiday in fifty years!

I should never have told Gladys about that successful interview at the bank. I'm slippin', — should've told her we were on the brink of ruination as usual, instead of shootin' m' mouth off. Serves me right.

She has it all worked out, — it's a week in Corfu, the Hotel Gigantos. Fly there, no bother, only £250 each, half-board. It could be terminal!

I won't be able to sleep, it'll be too hot, the mosquitoes are eating me already, the moussaka is at war with the swordfish, a dago policemen has thrown me in gaol for ogling topless nymphets on the beach, the plane has crashed on the wastelands beyond Tow Law, and I'm the only survivor!

Gladys snores all night, with a smile on her face.

Monday March 2nd

Fertiliser arrives mid-morning. The driver sits in his cab and reads the *Sun* (it's surprising how long it takes him to read the *Sun*) while I nervously remove the pallets with the fork

lift. There's no doubt about it, I'm a dog and stick man really, — anything more mechanically complicated than a hammer, and I'm ill at ease. There are so many knobs and handles on this machine I might as well be driving a moon rocket, — and I do well (I reckon) to smash only two pallets and hole a mere seven bags. My relief when it's all over is immense.

This total lack of sympathy and understanding with anything made of nuts and bolts must be hereditary. M'father was the same, — couldn't replace a broom shank let alone a spark plug. If a tractor conked out, he'd just climb down, look at it with digust, swear, kick it, then limp home to phone for a mechanic. (He always wore wellies, so his toes were in an awful state. Said he had footrot, but it wasn't, — we knew it was from kicking dead Fordsons.)

Tuesday March 3rd

Get the yowes in and give them their pre-lambing booster injection. They seem to be in an agitated mood, but maybe it's just my fettle. One jumps out of the pen like an antelope (assume she must be geld). Another stands by herself, coughing (assume she's preparing to die), and a third has developed mastitis (assume she'll have triplets).

Got the job finished by twelve, without seriously injuring any sheep, or myself, or kicking the dog.

I can remember when we gave the yowes nowt but food and water, — but that was 'yesterday', when everybody had the proper number of sheep. What was it m' father used to say? ... 'you can increase your sheep numbers as much as you like, — but they'll eventually just die back to what the land can carry. ...'

Gladys made pancakes at dinner time. I love them rolled up full of lemon juice, butter and sugar. I ate five before she got a couple for herself.

In the afternoon I took a walk around the whole estate, to see if we can travel on the land yet. The winter corn will be to top-dress and spray, and there's some spring barley to sow ... and then the swedes of course.

The winter barley looks not s' bad. There's one patchy field which we had to 'plunge' in, and it's *still* soggy, — maybe a kick in the roots with nitrogen will encourage it, and turn the yellow to green.

The ploughed land is drying slowly, but it won't cultivate yet. We need a few breezy days. The turnip field might plough, but it would probably be better to leave it, then plough, work and sow, all in one go.

I enjoyed the walk, birds twittering, crows carrying building materials to the rookery, — (no, that canna be right, — what do they say? — if you see a rook flyin' by himself, he's a crow ... and if you see a flock of crows on the wing, — they'll be rooks). Anyway spring is in the air ... I think.

Wednesday March 4th   *Ash Wednesday*

Today the sleet is blowing horizontally straight from Siberia.

Sometimes I think Mother Nature is the head of some terrorist organisation, dedicated to the disruption of farming life. She can certainly be a moody bitch. If it was 'Father Nature', things would be much more stable and predictable.

Which reminds me, — Doreen is in despair. Apparently she and Melvin have fallen out. Gladys is very sympathetic, blathering on about lovers' tiffs. Personally I thought Melvin was a twit, — he smiles too much, it's not normal. Anyway Doreen is going about like a gimmer with staggers, head down, just lyin' about, — neither use nor ornament. Gladys says she'll get over it, — it's just part of growing up.

Why is Gladys so reasonable with everybody else, and not with me?

Thursday March 5th

The farm is awash, that soggy barley field is a lake, with water fowl floating merrily on it. The yowes look miserable, the tups in the croft look miserable, even the bullocks in the hemmel look miserable, and of course Doreen looks miserable (thank God she's away all day).

Furthermore the entire village is miserable, because I forgot to turn the banger off, and it boomed all night. Everybody (it seems) has been on the phone this morning (some of them not at all polite) and sure enough when I went for the papers, Mrs Simpson looked decidedly bleary. But then rumour has it she leads a very full nocturnal life anyway, — so it might not have been *my* banger that kept her awake.

Finished preparing the maternity ward (lambing lift-off is seventeen days and counting ...). Went to the vet's and topped up with medical supplies ... and bought a new spade.

Friday March 6th

To the Swan at 8.30 pm.

Reports are coming in that Cecil Potts is having a moderate lambing (a lot of watery-mouth and staggers). He wouldn't admit to anything less than a triumph of course, — but Jake the postman has seen a pile of ex-lambs, and at least three 'late' yowes behind the hayshed. He reckons he even heard Cecil sobbing pitifully in the byre, — but that might be what Florrie calls 'pathetic licence'.

We also discussed (quietly) the sex life of the widow Simpson. Charlie reckons there's a different car parked outside her house every night. He said he'd seen Davy Scott's car there, but Davy just smiled. We drew our own conclusions. He's a bit of a lad is Davy. He's one of those canny doe-eyed, sleepy blokes that women seem to fancy. They maybe want to 'mother' 'im, — but I suspect Davy's after more than hot cocoa and a cuddle.

Saturday March 7th

There's a howling gale. The lake is receding, the wildfowl have migrated, and the yowes look marginally less miserable.

Willie played rugby against what he called 'some poncy school' in the city. The bus picked him up at eleven o'clock. He came back twelve hours later with a limp, a dislocated finger and a fat eye. Gladys and I were watching Match of the Day when we heard him singing songs in the kitchen. We noted that his vocabulary has been enlarged as well as his eye. (With language like that, he could end up as a hill shepherd.)

Sunday March 8th

Woke up to find a bullock at the front door. His mates have cultivated the lawn (it looks like the Somme) and then gone elsewhere.

How the hell did they get out? I'd tied the gate with barbed wire after the last escape, when they ended up in the churchyard. This time the rising muck had apparently helped them to ease the gate off the crooks.

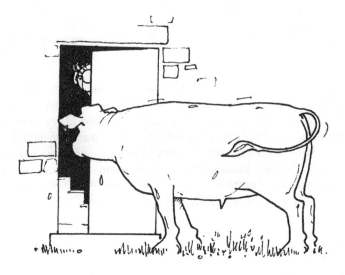

Fortunately they hadn't gone far, — we found them a mile up the road, where they had a Volvo estate surrounded. They were licking the windows, while two petrified old dears with a border terrier (yapping bravely) cowered inside. No damage to the car, just a slavery mess. We got the animals back home easily enough, — Willie in front making a noise like a bag of high-protein nuts, Sweep and me behind, having a sneaky cigarette. (Me that is, not Sweep.)

Monday March 9th

Tried to repair the garden with a graip and a barrowload of soil, and took a load of muck out of the hemmel gateway.

It's windy again, things are drying up. Cecil Potts has started top-dressing winter corn. Charlie says he's making a right mess, wheel marks up and down the field, — but it's still disturbing to see somebody else getting on with spring work before you. Charlie and I spent ten minutes on the phone convincing ourselves it was far too early — the land was still wet — and finally agreed that Cecil didn't know what he was doing.

After dark I got the fertiliser spinner hitched on the tractor.

Tuesday March 10th

Charlie (the swine) is top dressing winter barley! He says it's as dry as a bone, not a mark, — perfect.

I take another walk over *our* barley, and lose a wellie, — sucked off into a wet hole.

Cecil is cultivating land!

Wednesday March 11th

Six bullocks to the mart, there's nobody there, — they're all putting fertiliser on, or working land. I get a canny trade, but still don't feel happy. Have a quick sandwich in the canteen, forget to give any luck money, and hurry home to look at the barley field again. Are they all mad, or have I got the wettest farm in England?

At dusk I can hear tractors howling all around. Everybody is working except me. I can't stand it. Go to the pub (I never go to the pub on a Wednesday.) It's empty, except for Mrs Simpson and an amorous insurance salesman.

Thursday March 12th

It's too much. Cecil has a corn drill in the field. (I can see it from our hemmel roof.) Charlie is roaring up and down spinning nitrogen. Even sleepy Davy Scott, who is always way behind everybody else (makes hay in September, combining at Christmas), even he has his old set of discs rattling along. The sun is shining, the wind is blowing, and I feel terrible, m' guts are rumbling, — it's the nerves y' know.

Take another walk over the barley field and the ploughing (there's just a bit of white soil on the furrow tops now, — and I keep the wellies on this time).

At tea time Gladys comes back from the supermarket and asks why I'm not out on the tractor. 'Everybody else is,' she says. I consider killing her, but go for another walk instead.

Doreen comes back from the Poly and asks why I'm not out on the tractor. 'Everybody else seems to be,' she says. (I preferred her when she was miserable and silent.)

Willie comes in from a rugby training session, and says half the squad were missing, 'cos all the farmers' sons were helping their dads to sow corn.

After supper I load the trailer with fertiliser.

Friday March 13th

It's raining! Somebody up there doesn't like me.

We have a field of wheat showing signs of slug damage. It's thin in places, but it should be OK (I think). Old Fenwick Forster used to reckon that if you were worried about a poor crop of winter wheat, you should sit on the gate and throw your hat into the field. If the hat landed on a plant, then there was plenty there for a decent crop. It doesn't sound very scientific, but Fenwick never took much harm.

I began top-dressing mid afternoon with double wheels on. Anybody who tells me they aren't making a mark on their fields are either liars, or they're using a helicopter.

To the pub after supper, a little happier, — until Charlie says he'll be sowing corn tomorrow, and Davy announces he's got twenty acres in already!

Saturday March 14th

Finished top-dressing the winter barley by night time, — about two bags to the acre (don't ask me how many kilos that might be to the hectare, — that's EEC language) all I know is it's nearly seventy units. The same again (maybe a bit more, — if everybody else does) will go on in a month's time. Cecil Potts puts 200 units on, sprays everything seven times with complicated cocktails, and always ends up with four tons to the acre (or so he says). Maybe he needs it!

Gordon is working land for me. It's not too bad, — we'll go through it once, then leave it for a couple of days.

Charlie has a corn drill going. (Maybe we won't leave it for a couple of days.)

Phone Roger the contractor, and book his seed drill. He says he'll come over when he's finished at Charlie's, probably at tea-time tomorrow. There'll be no lazy Sunday this week.

Doreen is out with a new mystery boyfriend.

Sunday March 15th

I have the trailer loaded with seed and fertiliser before dawn. Hurriedly feed the livestock. One yow has lambed twins, and another is showing the familiar signs of impending motherhood. They would, wouldn't they? They're not due for another week yet!

Dragged a bewildered Willie out of bed, poured some tea down his throat, and stuck him on the tractor to cultivate. He went off like a drunk duck, wandering all over the place. But by nine o'clock he'd woken up, found Radio One, and the heater, — and was under control (more or less).

Back to the yowes. The twins are up and sucking, but the other one seems to have changed her mind, — nothin'. Carry the twins home to the hayshed, — food and water for the yow.

Check on Willie, who is now pretending to be Nelson Piquet. The revs and the radio are at maximum levels, the cultivator is in full flight, stotting along behind him. The whole combination is in a cloud of black smoke, and will undoubtedly disintegrate if I don't reach him in a matter of seconds. I have several cardiacs while running across the ploughed field to

give him a breathless bollocking. I need a cigarette, but have to wait till I'm back with the yowes before I dare light up.

She's had triplets, one the size of an undernourished hamster. He has no chance. Take him back to the maternal care of Gladys, who christens him Womble II, and asks if all the lambs are likely to be this big.

I relieve Willie (who is now bored to death in third gear) and have the field ready to sow by tea-time, — but there's no corn drill in sight.

Charlie's wife, Hilda, phones with the message that there's been a breakdown, and Roger won't get to our place till tomorrow.

Monday March 16th

IT'S CHUCKING IT DOWN! The cultivated field is like porridge, — I am a very unhappy peasant!

My disposition is not improved when I discover one of the twins (which in fact began life as a triplet) is now a single, — his brother having apparently been used as a mattress by mother. Mother appears to be less than distraught at this tragedy, so I give her a bloody good hammerin' (and break m' stick).

Womble II is reintroduced to mother, who fails to recognise the wretched thing. She will have to be 'persuaded', and is tied up in the byre, surrounded by hurdles, given another hammerin', and the perplexed Womble shoved in for a suck. He has no idea. This could be a long painful exercise for all of us.

Gladys informs me that Doreen's new bloke is none other than the son and heir of Cecil Potts the genius. That's all we need!

Tuesday March 17th

She's gone and sat on her other lamb! Now all she's got left is Womble, and she isn't very chuffed with him. So, what we've got is a great big super-fit mule yow, with enough milk for Elizabether Taylor to have a bath in, — supporting one little rat, who hasn't discovered where the tits are!

And it's still raining. Fetch the rest of the yowes into the back field in case any more decide to 'entertain' us.

Wednesday March 18th

Took the last of the fat bullocks to the mart. Ten others left, including the pot-bellied ringworm thing, and the alcoholic one. They'll get fat off the grass.

It's a good trade (a 'flyer' in fact) but the canteen chat was depressing, because most of the other blokes had managed to get at least one field of spring corn sown.

Charlie had been 'caught' by the rain (and the breakdown) and still had about three acres of butts to sow.

Son of Cecil phoned at night to talk to Doreen. He's called ... (wait for it) ... he's called Wayne! I couldn't believe it, fell about, — Doreen said I was a philistine (whatever that is) and then talked to him for an hour and a half. Thank god he phoned *her*, — Cecil will get the bill.

Wayne!

Thursday March 19th

It's still raining, well ... drizzling.

I think Womble has found a tit at last. He's as full as a little fat rat can be this morning, — but we'll leave mother tied up for another day to make sure.

The rest of the flock seem ominously quiet, — is this the lull before armageddon?

Friday March 20th

Fair, — just, — one of those dull, grey days, — a growing day, a lambing day perhaps, — but no, nothing emerges.

Took Willie to YFC dance, and went on to the Swan, where a heated discussion ranged from Mrs Simpson's interesting love life, to the Common Market cash crisis. Percy reckons the nation's morals have gone to the dogs, and we should bring back the birch. (I'm not sure if this has anything to do with Mrs Simpson's behaviour or not, but by all reports she'd probably quite enjoy it.) Percy also feels that the Common Market should be scrapped, and the man who was Minister of Agriculture after the war (Williams) should be reinstated immediately. Somebody pointed out that the bloke was sadly long dead, — but Percy reckoned he would still do a better job than those 'froggie fairies' in Brussels who should all be put in plastic bags and thrown into the Rhine.

I left to pick Willie up at quarter to twelve. No sign of Willie outside, and to my surprise he was actually dancing. It was quite embarrassing — there's little doubt that if he and Tracey had been performing, whatever it was they were performing, horizontally instead of vertically, — they would've been arrested (and birched by Percy). The boy has no shame!

Saturday March 21st

They're off! Two pairs, a single, three others who look as if they might do the trick before dinner, and an anorexic thing with staggers.

I got a dose of calcium into her, just before she died.

Move the pair and the single, iodine onto navels, little yellow pills down their necks, and by the time I get back there's triplets (but one's a gonna). No worry, twins will do nicely.

Two more sets of twins by noon, and one more yow scratching the ground over in the far corner. By night time we have twelve lambs from seven yowes, and one deceased. So far so good.

The bad news is Doreen has invited Wayne for tea tommorow. Gladys thinks this could be serious. It's serious right enough, who wants a son-in-law called Wayne? In fact who who wants a son-in-law? — a wedding will cost me an arm and both legs!

Had a late look at the sheep, and to bed at 12.30.

Sunday March 22nd

A good breezy day. We might get back to the corn sowing in a day or two, — maybe a quick run through with the cultivator first.

This was the official beginning of the lambing, so I leapt out of bed (alright then, crawled) at five o'clock. Nothing. Back and forward to the field all day, — nowt!

Wayne comes for his tea in daddy's Range Rover. He's about eight feet high, and built like a fork shank. He is also incredibly clean, and smells like Mrs Simpson. Worse still, he's wearing a suit!

I think I preferred Melvin.

Still nothing happening in the lambing field.

Monday March 23rd

They've started again (the yowes I mean) — but I haven't time to mess about with them. Roger's drill will be here at twelve, — so Gladys takes over the lambing, while I work the land.

In fact Roger arrives at 11.30 rarin' to go, so I have to stop cultivating and bring the seed and fertiliser into the field for him. I forgot about the bust drain along the headland, and sank to the axles. Much swearing and cursing before Roger pulls me out. He gets started at one o'clock, and I get the seedbed finished by half past, just as Gladys comes to report a purple-tongued lamb hanging out of a very mobile gimmer.

The gimmer is reluctant to be caught but eventually after three missed rugby tackles (once upon a time nothin' got past alive) and a lot of spitting and coughing and wheezing, I manage to grab the bitch.

The lamb is still alive, leg back, the size of a calf, and it takes half an hour to rearrange the animal before he'll come out. He'll make it, I think, — but we were just in time. He's got a head like Terry Wogan and a tongue like a stair carpet, but he's breathing. I squirt some milk down his throat, and leave him to work it out for himself. He's probably thinking that if entry into this world is so difficult, membership must be pretty exclusive (huh).

It's a fine quiet night, and if anything dies, it's just not trying.

Tuesday March 24th

Roger finished sowing last night. There are empty bags flying from here to the coast in a fresh west wind.

Meanwhile back in the lambing field things are going much too well, — a steady supply of twins, and no deaths. I'm a worried man.

Wednesday March 25th

Roger and his agricultural army came back to deal with the turnip land. One giant green monster with a five-furrow reversible started ploughing before breakfast, and by ten o'clock another massive machine with discs (that conveniently folded up to squeeze through gateways) was carving the land to a tilth. By dinner time the drill was here again, and before darkness fell, twenty acres had been sown and peace restored.

I have my reservations about these modern mechanical dinosaurs trundling over the fields. I worry about the drains, and the soil structure, but I haven't time to catch demented pregnant sheep, *and* sow corn, − so at times like this there's little alternative. And the great advantage of a contractor like Roger, is that *he* buys the sophisticated gear, and hopefully puts an expert in the cab. There's no way we could justify such machinery, or the bloke to drive it, for just a few days work in the year. This is a one-man band, and the only tune we can play is called 'survival'.

Thursday March 26th

What a glorious day, − warm, quiet, dry. (I can hear the traffic on the motorway miles away.) There's a pale spring sun, a fabulous day for lambs to be born. A magic day to wander round the field and count the new arrivals. A fine farming day, − you can almost hear the grass growing, the seed barley bursting. (The bullocks can smell it all, and they're restless to be out.)

Of course there are no lambs born on a day like this. It's just like the stupid sods, isn't it? Give them ideal conditions, and they just sit about sunbathin', − waiting for a blizzard!

Gladys says I'm lambing in my sleep. She claims I spent most of last night swearing at Sweep, and trying to get twins out of a pillow.

Friday March 27th

I heard the wind and rain in the middle of the night. Funny, − I can sleep through Gladys's symphonies, the telephone, probably a very clumsy burglar, − but rain beating on the window has me awake and alert in no time. (Well 'alert' is maybe an exaggeration.) Anyway I immediately pictured the yowes all lambing at once in a rapidly rising flood, and leapt out of bed at 3.00 am. Into wellies and leggin's, grabbed a torch, and out into the black wet night, full of apprehension.

Sure enough several had chosen this ideal opportunity to give birth.

Question: What to do first?

Answer: Find the lambs that are alive, and try to keep them that way.

Around the perimeter of the field I find three pairs, more or less on their feet, and a yow lying on her side, straining to unload *her* burden into the clarty mess she had already created. I don't know how long she'd been working at it, but there was no sign of feet or noses. Perhaps the lambs had taken a quick look outside, and decided it was more comfortable where they were.

Carried the twins into the byre and luckily the mothers followed quite well. Went back to find a three lambed in the middle of the field (the bleakest possible spot) one of them alright, two dodgy. Carried them inside, left the good 'n' with the yow, and took the other two into the kitchen (god bless the Aga).

By now the straining yow has a single (more to come I think). Grabbed the slippery thing, and persuaded mother to come inside.

So it's 5.30 a.m. I'm soaked to the Y-fronts. Tea and a fag. It's too much, − I'm smoking again!

Saturday March 28th

All the yowes into the sheds until this weather becomes more reasonable.

Yesterday we had seven pairs, a three and a single. One of the kitchen casualties died, but the rest should be alright. Two more twins overnight.

I'm not sure I can keep this bloody diary going throughout a lambing, — it's a very stressful period, not much sleep, irregular feed times, and damp knees.

The highlight today was a magnificent two-crop mule who, with no apparent effort, lambed three 'gerbils' straight into a ditch full of water, and then ran off to the far end of the field to bleat pathetically. The main reason why her rubbishy offspring didn't follow was because the poor things couldn't swim.

I managed to catch the mother when I brought the rest of the flock inside at the darkening, and gave her a damn good hidin'. She's not getting away with this, — it's a set-on situation.

Sunday March 29th   *Mothers Day*

I didn't know there was a special day for Mothers. As far as I'm concerned just now, — every bleedin' day is mother's day, and the mothers are all yowes! Nevertheless both Doreen and Willie produced sentimental cards, and they even bought a bunch of daffodils. Gladys was overwhelmed, and went about with a silly grin most of the day. Certainly the flowers will be left on the sideboard until they look like dead wickens!

That's the thing about kids, — just when you're convinced they're a thoughtless selfish pain they go and do something quite nice. It's all very confusing. However, we've got a lot to be thankful for. So far they haven't sniffed glue (as far as I know) or mugged some old lady . . . or joined the young Conservatives.

A yow has lambed while lying on her back, leaving her new-born son wandering about in a bewildered state. Instinct presumably told him mother's tits would be hanging down, — to find them sticking up in the air must be a surprise.

Willie is a big help today. What's more I'm pleased to see him join in the spirit of the occasion. He was carrying a pair into the byre, yow following uncertainly behind, when right at the door she ran away. Willie drops the lambs, and stands there, — the language is extravagent to say the least. I ask myself, does he know what these words mean? Where did he get them from? What would Tracey think if she could hear him now?

She can't, but Gladys can, and the poor lad gets a hard clip along the lug when he goes in for his tea. (End of Mother's Day celebrations.) Personally I'm not really bothered at all. In fact I'd say he was showing distinct signs of growing into a proper bad-tempered chauvinist peasant, — it's quite gratifying.

Triplets just before dark. I relieve the lady of one, and begin the persuasion procedure with yesterday's ditch disaster.

Monday March 30th

I've skinned one of the drowned lambs, and fitted out the triplet with a new overcoat. It's a very bad fit, — his inside

leg measurement is probably twice that of the last owner of the coat, and the poor thing can hardly breathe. His foster mother doesn't seem impressed either, and is reluctant to feed him. She will have to be 'encouraged'.

Tuesday March 31st

Back-body out, — or if you prefer it, — a prolapse. (Yes I'm aware it may be an unattractive subject, but peasants can't be sensitive little Victorian violets at a time like this. If you prefer to read Barbara Cartland, — get on with it).

Fixed her up, and began discing land for bagies (swedes).

Gone are the days when we sowed the seed in drills, then scarified and singled, pulled and chopped (I've still got the scars). Now we prepare a good seed bed, mix in the pre-emergent weed killer, and precision sow one plant every few inches.

Meanwhile the pigeons wait patiently, just out of sight.

Check the sheep before dark. We now have a fair number of ewes and lambs on the seeds, — two short of twins. The singles end up on the old grass with very little extra food (serves 'em right).

The yow with the set-on lamb is not being co-operative.

April

Wednesday April 1st

April fool's Day. Thankfully the family know better than to play practical jokes during the lambing, — the sheep however, are unaware of the dangers.

We have some trouble with a hogg. Hoggs, of course, have no previous experience of motherhood. When they produce their 'little bundle of joy', they are sometimes so astonished that they fail to recognise the creature as one of their own species, let alone their own flesh and blood. Occasionally they panic and run off at great speed into the middle distance when confronted with a strange bleating thing from inner space.

Our hogg had deposited her wee cross-Suffolk surprise while calmly grazing and, when the unfortunate lamb fell to earth, mother looked round and fled.

When I eventually caught her and fastened her up in the byre, she still didn't want to know, and the stupid lamb wouldn't suck, — even when I cowped the hogg upside down and stuffed the self-service equipment in his mouth.

How do you remain sane in the face of such awkwardness? I wasted an hour before I got milk into that lamb. Mother was still unconvinced about the merits of maternity — but she will be, if it takes me all week! My knuckles are sore from punching the bitch.

Cultivated the turnip land for the second time, and spun on fertiliser. Another pass should do the trick. It's been ploughed all winter and is breaking down well.

Thursday April 2nd

It's raining, — that's the end of land work for a while ...
ah well, the best-laid plans of mice and men. ...

Left Gladys in charge of the lambing for a few hours and
went with Charlie to a store cattle sale.

They were on fire, ridiculous, couldn't possibly be worth
the money, — the world's gone mad, how could anybody
afford such prices? There couldn't be any profit in them.

I bought twelve little bullocks.

Back home Gladys has had a serious chat with the hogg,
and it's allowing the lamb to suck. The woman's amazing.
Another day of aggravation and I would've felled the crazy
thing.

The set-on lamb with the small overcoat is still fighting for
his supper, — he's a game thing. Mother hasn't stopped
butting him yet, so I tie her up by the neck so she can't turn
round when he dives in.

Friday April 3rd

The bullocks look even dearer at home, — they've shrunk
since leaving the mart. Their only redeeming feature is that
they're basically the right shape. They've been half-starved,
muck stuck to their bellies, but they should improve and grow
into something useful ... some day.

Willie and I worm them after he comes home from school,
and kick them out into a sheltered field.

Wayne comes to pick up Doreen. He tells me Cecil has a
210 per cent lambing with no deaths (oh yeah?).

Willie is out on the town as well. Gladys has to collect him
tonight, 'cos I'm not leaving the lambing to go groping among
steamy cars in the early hours for passionate William.

Would you believe it, — Gladys says when she went for
him he was actually waiting at the door, all smiles and full
of thankyous. The devious little devil. ...

Saturday April 4th

The land has dried out enough to rattle through with the cultivator again. Phone Roger, — he'll sow the turnips on Monday.

We have a sick yow. (So what's remarkable about that I hear you ask.)

You're right, nothin', — wouldn't have mentioned it really, but it was a dull day. All we had was a big numb single with the IQ of a fencing post who goes bleating about after every other yow, upsetting the whole flock. It's a poor lamb who doesn't know his own mother. Also two pairs with about as much milk as a breeze block between them, a three from old back-body (I'm amazed they got out) and a gimmer who's convinced she's lambed (but hasn't) and licks every new lamb.

Tracey came for tea. I suggested Gladys had a little womanly talk with the lass, but she says Tracey is very mature for her age. (I'll drink to that!)

Sunday April 5th

The townies are out in force today. Leaving behind their semi-detached struggle in suburbia, they flee to the wide-open spaces of the countryside, and then park three deep in the lay-by, halfway between here and the village. (In fact it's just a bit of wide verge where the council tips gravel for future road repairs.) There they produce deck chairs and settle down side by side, the fellas reading the *News of the World*, and the women knitting, while their brats slide down the gravel mountain.

One family, obviously attracted by my colourful language, came to watch while I lambed a mixed-up set of triplets. Legs and heads all over the place, none of them connected. Took me half an hour to sort it all out, and by then the yow, the lambs and m'self were sick of the operation. However they seem alright. Gave mother a penicillin jab, and went away feeling like James Herriot.

God knows what the townies made of it all, — they seemed quite impressed, a lot of oohs and ahs.

Monday April 6th

Roger sows the turnips by dinner time. They should grow, there's plenty moisture there.

Willie goes back to school. I'll miss 'im, he's been a big help in the lambing field and on the tractor over the past couple of weeks. Of course he has to be dragged out of his bed. It's like raising the Titanic some mornings, but once he's actually awake (it takes about an hour) he's quite useful. I wouldn't tell him this, naturally, or the little waster will get delusions of adequacy, — but I might give him an extra quid or two at the weekend.

Spent the afternoon ringing and tailing lambs, — I try to do this job as soon as possible after they're born (in fact I'm breaking the law, castrating some of these older ones) — but I missed a few while we were working land. The tup lambs don't like it much of course, — writhing about on the ground,

— but it's maybe more civilised than the auld-fashioned method. M' father used the knife and his teeth (until he got false ones).

Tuesday April 7th

Dead yow. I thought things were going too well.

What did she die of? Who knows. I sometimes think it's their life's ambition to drop dead for no apparent reason. I wouldn't be surprised if they had a committee meeting to decide whose turn it was this week. They reckon a sheep's worst enemy is another sheep, — but I suspect it could be the shepherd.

There's another one looking dodgy, so I phone for Robbie the vet who stuffs her full of everything he's got in the boot of his car. I conclude he isn't very sure what's the matter with the creature, so he tries to cover as many alternatives as possible, at considerable cost to me. Still, if we can get a live lamb out of her it'll be worth it.

We had another fatality today, but I didn't mention it to Robbie. Remember the yow I had tied up by the neck? the one who wouldn't take the lamb with the skin on? Well she committed suicide, or maybe I had the rope too tight, — anyway she's a gonna, hanged!

I expect the lamb is delighted to be out of his overcoat. He's a pet now, one of Willie's unique flock.

Wednesday April 8th

There's a lamb short on the seeds. (Has he been rustled? I doubt it.) I walk the dyke back, look in all the hollows, behind trees, in the ditch, — but he's gone. Sweep runs about sniffing thistles, peeing everywhere, an anxious expression on his face. I'd rather find the lamb dead, half-eaten by a fox, eyes pecked out by the crows, than not find him at all. Predictably his 'devoted' mother is unconcerned, apparently unaware that half her family is missing!

I can't waste all day searching, so move on reluctantly to feed the rest of the sheep, check the cattle, and back to the maternity ward.

'Dodgy' has rid herself of two dead lambs. It's all part of life's rich tapestry, or so I tell m'self. You can't win 'em all, nobody promised it would be easy. Give the old thing another shot of 'Robbie's elixir of life'. She's got oceans of milk.

Should we try it? Why not? Nowt to lose! Rub the dead lambs all over the bewildered pet, and stick him in beside her. She thinks he's lovely, takes him no bother!

Thursday April 9th

Return of the prodigal lamb!

I can only assume the stupid thing had got through among Charlie's sheep next door, and fallen asleep. When he woke up hungry, he would have run about looking for his mother. I expect Charlie's yowes soon sent him packing, so eventually he ended up at the fence bleating for his supper, at which point his mother must've remembered she once had a pair.

That's the good news. The bad news is, 'Dodgy' is dead. Stiff as a board, gone to the great worm-free heaven, happy at last, — the bitch! This is especially unlucky for the poor lamb we set on yesterday. He must be very disillusioned about life. He's had three mothers already, and they've all left him orphaned!

However the lambing is winding down, — the annual festival when a peasant pits his wits and patience agin the massed awkwardness of pregnant sheep is drawing to a close. Only twelve left altogether, and chances are a couple of them will be geld, — so the twenty-four hour vigil in the maternity ward is nearly over.

Friday April 10th

Yow with mastitis, dragging a back leg. (She has a pair.) One side of the undercarriage is inoperative already, one lamb is struggling to survive, and his brother is unwilling to share his side.

Penicillin, and hope for the best.

Top dressing winter corn again, got it about right I think. I don't want any 'misses' or overlapping dark strips (particularly next to the road).

Of course the experts tell me to put more on, and use a straw stiffener for maximum yield, — some of them tell me to split the application, others say top-dress all in one go. They also point out a multitude of diseases, and encourage me to spray every other day. Then along comes a conservationist bloke who tells me not to spray at all, because I'm poisoning the lesser spotted Bolivian two-toed moth or some such endangered creature. They tell me to use LESS nitrogen 'cos I'm killing all the fish in the river, and the politicians reckon I'm just too bloody clever, and shouldn't be growing so much corn anyway! What do you do?

Saturday April 11th

The mastitis lady is very poorly, — take one lamb away to be bottle fed.

Willie and I get the rest of the girls in for a final pregnancy test. One gimmer and three old ewes (one of them very lean) obviously have no plans to have a family. So I've fed the selfish sods all winter for nowt (in future we might consider a scan). Anyway they will have to go, I'm not keeping them for another year, — they've made no effort to keep me!

After last night's production (two twins and a single), that leaves five to lamb (one a long way off) but we must resist the temptation to work out a lambing average at this stage.

In any case the only honest equation is lambs sold from yowes tupped, — but nobody I know does it that way. Most peasants talk in terms of all living lambs (including pets, rubbish, half-dead things) from the number of ewes that are actually running with lambs at the end of the operation. This way you can still afford to loose a couple of ewes and actually increase your average a little. Yes, we know it's a bit of a con-trick, but if you couldn't fool yourself occasionally you'd go crackers!

Wayne comes to take Doreen to the pictures, and says Cecil is now claiming a 215 per cent lambing. He finished a week ago, and it's still rising. Do we assume he's lost a few yowes?

Sunday April 12th

Mastitis gives up. We now have four pets, and Willie is given the responsibility of feeding them twice a day in return for one of them. He picks the best 'n', and names it Billy Beaumont.

The farming programme is concerned with 'set-aside' schemes. Will I be able to set-aside the whole farm I wonder, and pick up a hundred quid an acre for doin' nowt, and spend my time leading smiling parties of townies through the nettles and thistles and buttercups and wickens, that bloom in my abandoned wilderness? . . . and play golf in m' spare time? Sounds attractive!

Monday April 13th

Gave Charlie a hand to castrate calves. This is a dangerous job, — carelessness while approaching a worried bull calf from behind, armed with a Burdizzo, can lead to the wrong animal having his masculinity shattered.

Charlie never gets properly organised for this task. It's all done with spare gates and baler twine, and sure enough we always have at least one over-excited beast, urged on by blar-

ing colleagues, who smashes through everything, sending bits of wood, clarts, skin, wellies and petrified peasants flying in all directions. All he's trying to do is avoid a very nasty operation, and I suppose you can hardly blame him for that.

On this occasion we managed well enough until the very last victim. Of course we both got the odd kick in the knee, we were both covered in muck, we'd smashed a few sticks, there was blood about (some of it mine), but on the whole the performance had gone fairly successfully. Then this cross-Charolais maniac wouldn't go into the crush (well you know what Frenchmen can be like). Twisted tail, a flood of obscenities, a good hammerin', even a tempting bag of barley meal, nothing would persuade the bugger to go in. He bucked and kicked, bellowed and snorted, — knocked Charlie into the dipping pens, smashed through a gate, leapt over the stackyard wall, and roared off down the road towards the village.

The cows, locked up in the hemmel until their sons had been dealt with, all became very anxious, broke open the gate, and stampeded off in pursuit of the runaway calf. Charlie and

I followed in the pick-up, and just got past them at the War Memorial, but by then they'd plunged through a few gardens, and left a trail of dark green stuff for over a mile.

The widow Simpson was hanging out some interesting washing when they swept through her place. Fortunately Gordon was having a day off from mart duties, and Jake the postman appeared from his coffee break at the pub, and the four of us eventually got the herd home again. By then the reluctant bull calf was thoroughly guffed (we all were) and he walked straight into the crush apparently resigned to his fate.

Tuesday April 14th

Finished top-dressing the winter corn, putting a bit extra onto the south field of winter wheat, the one with the slug damage. Managed to reverse the spinner into the side of the trailer (too much hurry) and bent the hopper. This machine is only two years old, and already looks as if it's been to the Lebanon.

Wednesday April 15th

Today Charlie and I went off to a sale in search of cheap cattle. We discover cattle are an endangered species. I managed to buy the first one into the ring (a Friesian cross-Hereford bullock) before the audience had settled down, but after that the trade went berserk.

Humphrey Smith, who is a farmer, cum dealer, cum scoundrel, bid at everything. He was a proper nuisance, just pushed the prices up, then generally dropped out. He started most lots where they should've finished. I got two more while Humphrey was having his dinner. Charlie got nothing. (He did well.)

Thursday April 16th

As cattle are so expensive I decide to buy some more sheep instead. Perhaps a few young ewes bought now with lambs will save me a quid or two in the back-end, when I would normally be buying gimmers or hoggs.

So off we go again to look for gimmers and twins.

There is obviously a world shortage of these things as well!

That pest Humphrey is there again (he goes to seven marts a week, two on a Friday) and naturally he wants gimmers and twins today. I wonder what it would cost to have him assassinated?

Anyway I couldn't touch them, they were far too 'hot'. I got a bit desperate and very nearly ended up with a pen of antique Blackies with singles, all of them lame, − but luckily Humphrey came in with another bid, just as I was creeping out of the ring.

It was a long day for nothing, − just smoked too many fags, and didn't get a proper dinner either.

Friday April 17th   Good Friday

Three yowes left to lamb, one as lean as a crow, like Twiggy with a woolly cardigan on. She'll probably have triplets.

I began dosing and injecting lambs for worms and a few other problems. It's the same sort of philosophy as with the corn spraying ... it all costs a bomb, but you feel you must try to eliminate at least some of the cock-ups that would occur if you did nowt. Sometimes I feel farming is becoming too complicated for mere farmers.

Meanwhile the byways are full of townies out for a 'run' again in their shiny Sierras. Dad driving, as if piloting a 707, mother pointing out things of interest, Granny full of lunchtime sherry, burping in the back, the kids sick on a surfeit of smarties, and bored (they'd rather be watching an old James Bond film on the telly).

'Look', cries Mummy, 'some little lambies, with a nice old shepherd . . . '. And there over the hedge is someone like me, who has to stop beating the living daylights out of an eight-crop mule, who's just had a pair, but can't count beyond one.

Saturday April 18th

Gladys has bought a new fireside mat for some outrageous price. I could've bought two gimmers and twins for that money! We had a row about it. But she always wins, because even if I imagine I've won (by superior intellectual argument) she just stops making me cups of coffee, plonks down some lumpy mince for m' dinner (or even liver — I can't stand liver) and the potatoes are hardly cooked, and there's no puddin', and she vacuums the entire house (twice) while I'm trying to have a kip. It's impossible, I'm no match for the woman. Eventually I'm forced to admit I was being unreasonable!

Meanwhile Willie is playing in a 'sevens' competition, the last event of the rugby season. There's a dance afterwards, and I have to collect him as usual. This is never good news, but at least we get him home, — god knows where he'll end up when he's old enough to drive a car.

When I went to get him, he appeared in the headlights flat out on the grass beside the club house, with the angelic Tracey apparently nursing him back to health. For a moment I naively imagined he'd been injured in the match, or beaten up by skinheads, and Tracey might be giving him mouth-to-mouth resuscitation.

Wrong again. Gladys and I didn't go on like this, did we?

Sunday April 19th   *Easter Sunday*

A ewe gave birth to twins last night and is looking a little fragile. I bring her into the byre and dose her with Robbie's

elixir, — at which point she promptly collapses and dies.

What the hell was that for? ... a heart attack maybe? ... a brain haemorrhage? ... No it couldn't be a *brain* haemorrhage. A haemorrhage possibly, but not a *brain* haemorrhage. Ah well, two more pets for Willie to feed, — and the lambing average goes up.

Wayne comes for tea, sits grinning at Doreen like a labrador.

After tea Twiggy lambed. No, not triplets, not even a pair, — just a neolithic single, who's nearly as big as his mother. And she's so chuffed with him, she won't stand still long enough for him to suck. He's wobbling about bleating for his dinner, and she's following, trying to tell him that she's got it ready, but he can't find it. What we have here is a breakdown in communications. Shut them both in a very small pen to work it out.

This is a weekend when country folk shut the farmyard gate, and stay at home. The lanes are full of caravans and nomadic ramblers dressed in big boots and hairy knee socks.

Monday April 20th *Easter Monday*

Walter Williamson was a peasant like me once upon a time, but his wife ran off with a worm-drench rep one Wednesday while he was at the mart. He recovered from that eventually, but then his collie dog died, and he went all to pieces. Now he works for a spray company, and advises us ordinary mortals what concoction to use. He tells me something called leaf spot is likely to destroy the rape; mildew and rynchosporium will decimate the barley; yellow rust, septoria and eye spot will wither the wheat. He works it all out, sends in his sprayer, and I get a series of bills which will probably wither me. There's no escape. His machine is here today spraying winter corn, Bank Holiday or no Bank Holiday, − the conditions are right.

Meanwhile a check on the lambing reveals a crop of 168 per cent. The best we ever had was 170 per cent, − the worst 165 per cent. Cecil Potts would obviously consider this to be a disaster, and would quote his own extraordinary figures, together with a host of bewildering costings from his gross margins and variables, until he had convinced everybody that they were incompetent nincompoops bound for the workhouse (that's if they didn't know already not to take him too seriously).

Saw a swallow this afternoon, and put the wintered bullocks out of the hemmel into the front field.

Tuesday April 21st

Walter's man is still spraying in good calm conditions. I decide to undersow the wood field with grass seeds, using the fertiliser spinner.

This has always been a highly technical job, requiring immense skill and precision. For a start it takes me half a day to work out what kilos to hectares really means, then I measure out an acre, set the spinner at what the manual suggests, empty an acre's worth of seed into the hopper, climb aboard the tractor and, carefully selecting the appropriate gear and revs, — off we go.

It works quite well. At the end of the measured acre there's no more than a wee pickle of seed left in the bottom. It seems reasonable then to increase the flow just a teeny weeny bit. This proves to be unwise, because after sowing a quarter of the field I discover half the seed has gone. It seems reasonable therefore to marginally *decrease* the rate. Off we go again stopping later to check how we're doing, — only to discover (and panic) that I've sown virtually nothing over the last five acres. I sow the last five acres all over again, then readjust the damned machine for the rest of the field, finishing in a cold sweat, and in the dark, with enough seed for three acres to spare. It's always a complete mathematical cock-up, — but a year later y' canna tell.

Gladys is packing, and the holiday's a month away yet!

Wednesday April 22nd

Charlie has coccidiosis. Well not him personally, but his lambs have, according to Robbie the vet. Ye gods I thought only hens got that, — but I should've known, if hens can get it, sheep will insist on having it too.

Walked round all our sheep looking for coccidiosis. I don't think we've got that, but they're nearly all lame.

Looked at our seven hens to see if *they* had coccidiosis, but you can hardly tell what our hens might have, — they're now so old and bald they could have anything. Once they were handsome Rhode Island reds, all plumage and arrogance, an egg-a-day birds, — now they're naked and off-white Rhode Islands, who stagger about the stackyard laying a very occasional egg, generally where nobody can find it. Maybe we should buy a new flock and eat this lot. Gladys is in agreement with this, but reckons they would have to be marinated for a week before they'd be edible.

Thursday April 23rd *St George's Day*

I had intended to deal with lame sheep today, but it's pouring!

So I end up at a mart, and buy ten cattle in a 'fast' trade from eight different sellers over a period of three hours while eating ten cigarettes. It was even harder work getting some luck money from the tight-fisted vendors.

Walked around the stock before supper in a drizzle, and discovered a lamb lying by himself looking as though he'd had a bad night at an Indian restaurant. Carried him home and pumped him full of everything I could find with an expiry date after world war two, and put him under a lamp in the shade.

Rambo the monster single has at least discovered where his future hangs, and is full as a barrel. Out he goes with the other singles.

Friday April 24th

Walter says we will have to wage war on the weevils next month, they're attacking rape crops down south already. Sounds like a bad science-fiction film. I have a certain sympathy with those conservationists. Here we are injecting everything that moves, and spraying everything that doesn't, — and I really haven't got a clue what I'm using. Has anybody, I wonder?

After yesterday's rain the swedes are emerging quite nicely, not many gaps. No doubt the pigeons will have noticed the crop, so I install the banger as a precaution and a deterrent.

Top-dressed the seeds. We'll have to remove the stock soon to allow for the annual hay-making heart attack come June.

Gladys is becoming very excited about the holiday, marking off the days on the calendar. She and Doreen are constantly discussing clothes. Willie observes that nobody wears clothes in Corfu.

I understand Corfu is well south of Darlington.

Saturday April 25th

We are invited to a party. Gladys accepted the invitation while I was distracted by sheep, and in no condition to argue rationally. I knew it would be a disaster.

The Pillicks have recently settled in the village, bought two old cottages, and transformed them into the sort of mini stately home you might see in *Good Housekeeping*. Tonight it was open to the public.

I didn't see the bedrooms of course, but Gladys had a sneaky look at one, and reckons it's like a tart's boudoir. How the hell does she know? She says there's mirrors on the ceiling, and a massive sloshy thing called a water bed. So you can lie on your back and watch the wife drown, I suppose. Gladys certainly would 'cos she sleeps with her mouth open.

The bathroom is so posh, shiny and marbly that normal folk like me feel positively guilty about using the loo (that is after you've worked out which thing *is* the loo). The kitchen is like an operating theatre, all built-in gadgets, clinically clean, no cleaky mats, no dirty cups, no sticky cough bottles, no wellies. A perished lamb wouldn't survive five minutes in there ... and neither would I.

The whole set-up must've cost a bloody fortune.

Once upon a time Charlie's old tractorman would've retired into one of those cottages, or Percy's mother, or Arthur's shepherd. But not now. They've got to end their days with some reluctant relation, or in an old people's home without a garden.

The Colonel doesn't help the situation either, because as soon as some old rural lodger dies, he puts the house up for sale, and a posse of well-heeled townies compete for their 'place in the country'. Before you can say 'upwardly mobile', the village is awash with green wellies, *Daily Telegraphs* and BMWs. What happens then? Well, the newcomers shop at the hypermarket where Daz is 2p cheaper and they cart Amanda and Nigel to a private prep school, so the village shop is knackered and the village school is empty. The pub goes all brass and plastic leather with microwave quiche, and the dart team dies. This is called progress!

Anyway you couldn't rabbit on like that at this party, there wasn't another peasant in sight. All the blokes wore suits and talked about the Footsie, pension schemes, wealth creation, and holidays in Gambia. Worse still nobody was smoking.

We were the first to leave at midnight, Gladys a bit silly after two dry martinis.

Forgot to turn the banger off on the turnips, but couldn't be bothered to go all the way down there in the dark.

Sunday April 26th

Jimmy and Edna Fletcher called in for tea on their way back from a long golfing weekend in Scotland.

Ye gods, that man's never at home. Has he won the pools, I wonder? Did his granny leave him a fortune? Or is it just because they've got rid of the kids now? Anyway they certainly lead a canny life, — fancy car, umpteen holidays every year. I don't understand it, it's a mystery. He's no cleverer than me, is he? (I saw him buy a pen of right bad dairy-bred bullocks a few weeks ago, and they'll die in debt for sure!) But maybe I'm just jealous.

However Edna wasn't entirely happy about this latest trip, she couldn't wait to tell Gladys.

Apparently Jimmy, Edna (and Edna's little Corgi dog, who goes everywhere with her) stayed at a right fancy hotel near a very exclusive golf club. It was one of those ancient Scottish clubs that claims to be the first in the world (there are several of them in Scotland) and they were both looking forward to playing the course.

Jimmy phoned up from the hotel, and booked the tee for nine o'clock on Friday morning. He didn't mention the dog. Just said there'd be two of them playing.

They arrived at the starter's box with five minutes to spare, Jimmy in tartan trousers, Edna in a tweed kilt, and the eager Corgi attached to her trolley by a tartan lead.

The old weather-beaten starter, whose undisputed rule had lasted over half a century, a man who had once actually shaken hands with Bobby Jones, and over the years had collected a considerable fortune in tips and bribes from rich American 'rabbits' and wealthy oriental 'thrashers', came slowly out of his little hut and surveyed the visitors with undisguised contempt.

'Mr Fletcher is it?' he asked, without taking the pipe out of his mouth. 'And Mrs Fletcher is it?'

'That's right,' said Jimmy cheerfully. 'We're off at nine o'clock.'

'Aye, that's as maybe,' grunted the old potentate. 'But I'm afraid it's no possible. ... '

'But I booked,' said Jimmy indignantly. Then it dawned on 'im ... 'Is it the dog?'

'No, no,' said the starter with a condescending snarl, 'your wee dog's most welcome ... but women (he pronounced it weeman) are not allowed on *this* course!'

Edna may never fully recover from the humiliation. In fact she nearly choked on a digestive while she was telling us about it.

Monday April 27th

Put some nitrogen on the spring barley.

I see in the shed that all the fertiliser has gone except for two rock-hard bags of 20.10.10, and all the seed corn, except for half a hundredweight chewed by mice. We also have ten bales of mouldy hay and about twenty burst bales of straw. This serves to remind me that now is the time when the overdraft is over the hill. All I've done over the last few weeks is spend, spend, spend. Nothing coming in but bills. We have what the experts would call a cash-flow problem, — but it doesn't take an expert to work out which way it's flowing.

Tuesday April 28th

Today is sheep's feet day, everything gets a pedicure. The lambs have scald, the yowes have outgrowing toe-nails, even the tups are limping about like wounded soldiers. Charlie calls in, and he's limping too, — a cow stood on his foot.

Getting the flock through the footbath is a swearing, punching, kicking, spitting, stick-waving pantomime. The lambs won't go unless hurled bodily into the bath. The yowes either stand stubbornly refusing to budge, or leap the whole length of the trough. I drag one old mule through, hoping

a few will follow, but no luck. I grab another awkward bitch, pull her halfway and am immediately overwhelmed by the rest of her mates. By the end of the operation I'm suffocated by fumes, covered in muck, eyes and nose streaming . . . and limping with a wonky knee.

Wednesday April 29th

All non-productive sheep to the mart. These include a tup who was rapidly pining away and would certainly have disappeared altogether in another week. Also a skeletal seven-crop mule, who luckily stood up long enough to be sold for a fiver which, bearing in mind she had to be carried into the horse box, carried from there to the mart pen, and carried from pen to ring, was miraculous. I suspect she expired before she left the mart. The geld gimmer and the two respectable ewes sold well. I clipped them before they left home, but not the tup and the old yow, who would've looked ridiculous without their clothes on.

Thursday April 30th

Gladys demands final payment for the holiday, — there's no escape now, she says triumphantly, once you've paid, you gotta go. Then, quickly realising that nothing is more likely to persuade me to opt out than a fait accompli, she adds 'it'll be good for you to do nothing for a day or two pet.'

A whole week doin' nowt?

Some lambs in the back field have orf. Phone Robbie who promises to send vaccine . . . 'there's a lot of it about,' he says.

May

Friday May 1st

Spend most of the day dosing scoured lambs and pushing them through the footbath again.

To the pub at night to cheer m'self up, only to find Charlie talking about coccidiosis (he's an expert on the problem now), Cecil telling everybody how he bought some cattle dirt cheap (so *he* says), Percy in the middle of another hanging speech (it's car park attendants this time, — he got a ticket last week), and Gordon forecasting the wettest summer in living memory. (He apparently saw a snail halfway up a tree, — presumably trying to avoid the impending flood.)

Saturday May 2nd

Spend the day fencing (i.e. mending fences). Just m'self and Sweep, with the tractor and trailer. Sweep sleeps most of the time under the trailer, while I bang in a post, nail on a rail, tighten up a wire. It's all designed to keep cattle and sheep (at best) in the right field, or (at worst) on the right farm.

Willie played cricket for the local team. He's hardly county standard, — bats at ten or eleven (depending on whether Gordon's playing) and fields on the boundary, because he's the only one who can throw the ball in from there. He scored a four, outside edge through the slips I think, and caught a skier at deep square leg.

Sunday May 3rd

A typical English Sunday afternoon in the country. Doreen is off with the bean-pole Wayne, Willie is away to ... well you never know with Willie, he just disappears, Gladys has gone for a walk up to Hilda's, and I have the house to m'self.

So I'm dozing nicely, full of Yorkshire puddin', when there's a knock at the door. For a while I pretend there's no one at home, — but there at the bleedin' window is this entire family peering in at me, and gesticulating about something. Turns out to be a car-load of worried townies, who tell me there's a wild bull out on the road, and he's terrifying some picnickers on the lay-by.

In fact it's one of Charlie's old suckler cows, a canny old cross-Hereford beast, and she walks back up the road no bother. The townies think I'm a hero.

Monday May 4th   *Bank Holiday*

Spend all day hiding most of our outdated (but interesting) machinery in the wood. Some of it is still there from the last time we had a visit from the Safety Inspector, and he's due tomorrow.

None of it looks particularly dangerous to me, but past experience suggests that this bloke views everything mechanical as a potential killer. He gets very excited at the sight of a naked reaper blade, or an uncovered PTO shaft, unguarded belts, cogs, chains, ladders with wobbly rungs, and all manner of normal everyday farming gear.

It's best if he just doesn't see it.

Tuesday May 5th

The nosy little bugger arrives at dinner time, so we're obliged to give him a cup of tea. You've got to be nice to those blokes, or they can make life very difficult.

'You don't seem to have much equipment,' he says look-ing at the empty shed.

'No,' says I, 'we're very poor, and can't afford all the fancy stuff, − we rely on Roger the contractor. ... '

He didn't like the barley grinder very much, especially when it burst into life as he switched on the light (must be a crossed wire somewhere) and he seemed to think someone's children were sure to come to the farm and drink the dregs from the old spray cans. He was a little nervous about the granary steps, particularly as three were missing, and the handrail came away from the wall when he grabbed it. He became quite upset when he saw the auger without a guard, and was giving me a fairly serious lecture on the dangers of such things, − when he disap-peared through the granary floor.

He said he'd draw up a comprehensive list of safety measures I must attend to, − then we carried him to his car and he went away.

Wednesday May 6th

Brought the 'scabby' sheep in from the back field and treated them for orf. It's a terrible disease and I was persuaded by Gladys to wear rubber gloves, and to be especially careful not to vaccinate m'self.

When I was finished she took all my good working clothes away to the wash, and made me have a Dettol bath before supper. I noticed nobody sat anywhere near me all night.

Thursday May 7th

Move the sheep from the seeds (hay ground). I could still do with a few more ewes and lambs. There's a farm stock sale next week, maybe we'll get some there, — but farm sales are often expensive.

Harry Thornton, who's a hill farmer living about fifteen miles west of here, has offered to sell me some sheep, but I don't fancy his variety. Harry's a great big hairy bloke, with two monstrous sons who only leave their mountain retreat to go to a sheep sale. He's a good stockman, from the old-fashioned school, but a pen of his wild fleet-footed hill yowes would never settle in a field down here, — they'd be off back to the high ground, through fences and over stone walls, as soon as they got out of the wagon.

I suspect all Harry's livestock have a homing device bred into them.

Friday May 8th

At dinner time I find Florrie (the lady who comes to help in the house) advising Gladys on her holiday wardrobe ... distinctly heard the word 'bikini', — Gladys in a bikini?

Florrie, who has never ventured beyond the river Tyne,

says she would never 'gan ower the watter', because of the language problem.

To the Swan after supper where Charlie is sympathetic about the holiday, reinforcing my worst fears, such as the risks in travelling 30,000 feet above the ground, and eating garlic for breakfast.

Percy reckons all foreign food is terminal, — Italian pasta, German sausage, French frogs, Welsh rarebit.

Saturday May 9th

Began mucking out the hemmel this morning, and quickly holed the water trough with the fore-end loader. Turned off the water, forgetting that this denies the farmhouse too. Gladys comes out to remind me.

We had another (minor) tragedy as well. This very smart gentleman stops me just as I'm driving out of the yard with a full load. 'Morning,' he says, 'I'm from the Inland Revenue.' I was so surprised I must've accidentally knocked the P.T.O. into gear, and he got covered in smelly stuff. I

apologised of course. He says he'll have to come back another day. Serves 'im right for sneakin' up on a Saturday.

Spread the muck on the old grass field next to the village, and the Pillicks predictably make quite a fuss about polluting the environment, and how the smell could ruin their weekend.

Davy Scott, Gordon and Mrs Simpson each ask for a load to be tipped in their gardens.

After dinner I go to watch Willie play cricket. He's bowled first ball and later drops a catch on the fence, while chatting up a well-endowed fifth former in a red jump suit. However, in spite of Willie's efforts our team wins.

The scores are: them, 47 all out; us, 49 for 9, thanks largely to our umpire (Arthur Thompson's dad) who would go down very well in Pakistan.

At night Wayne arrives and asks Gladys if he and Doreen can go on holiday together ... 'separate rooms of course,' he says (the creep). Doreen smiles coyly as if butter wouldn't melt in her mouth.

Gladys says she's sorry, but Doreen's father just wouldn't stand for it, — 'he's very old fashioned,' she says. And there's me sitting watching the telly and listening to all this stuff. She's right of course, but she doesn't have to make me out to be a geriatric killjoy.

And anyway, if Wayne's so bloody keen to have a holiday, perhaps he'd like to take my place in Corfu ... ?

Sunday May 10th

First game of golf this year, — a four-ball with Jimmy Fletcher (he of the European circuit) and a solicitor called Rodney (off a handicap of about six) against Arthur Thompson and m'self. It was the normal procedure, a couple of drinks before we started, then a ten-minute conference to negotiate strokes and stakes, — and away we went at one o'clock.

We were all square at the fifth, a good battle developing, but it was then that Arthur went to pieces. The solicitor had just holed a twenty-foot put which had upset Arthur, and on the next tee he went for a big drive and sliced it badly.

Now down the right-hand side of this fairway there's a caravan park, very sensibly protected against approaching slices by a high wire mesh fence. There was however one very small hole in this fence, and Arthur's ball went straight through it at the speed of light. It was a chance in a million. Happy caravaners sitting in the spring sunshine having lunch were forced to take evasive action as the old Dunlop 65 screamed past them and hit a sleeping border terrier.

Arthur was in no mood to give comfort to some unlucky hound though. Producing his *other* old Dunlop 65, he placed in on the tee, and smote with a terrible venom straight down the middle of the fairway.

Unfortunately, with the force of his swing, the three wood he'd used slipped out of his hands and flew like a whistling wayward scythe over the fence and in among the caravaners again. By now the somewhat nervous campers had just settled down after comforting the terrier, but were obliged to fall flat on their faces as this second missile came among them.

Give Rodney his due, as a solicitor, he thought he might be best equipped to placate the victims of this unprovoked attack, and off he went to talk about 'Acts of God' and 'diminished responsibility'.

But Arthur was never the same again, and we lost 4 and 5.

Monday May 11th

Warnings that nematodirus is rampant. Florrie informs me that even the Archers have been plagued with it, so we get all the lambs in and dose them again. Did sheep always get these problems? What did we dose them with years ago? What did we dose ourselves with years ago? — Castor oil on a Friday night, if I remember rightly.

While the sheep are in the pens one yow begins to shake all over, and her eyes glaze. I saw Willie behave like this one night at a disco, — but he lived. I'm not so confident about this beast. Her lambs, oblivious of what her problem might be, still dive in for a suck. They lift her back-end clean off the ground, and she continues to vibrate in mid air. By the time I've finished dosing, she's in a bad way. We put her and her lambs into the stable, and phone Robbie.

Tuesday May 12th

Robbie arrives to view vibrating yow at 8.45 am.

Sadly she ceased to vibrate a quarter of an hour earlier. Robbie has a cup of tea, and says he won't charge me for the visit, 'cos he's going on to Arthur's anyway.

Sell the pet lambs to the Pillicks, who are now into what they call the 'rural scene', and already have a goat, three hamsters, four hens, a rabbit, and a fat pony for their spherical daughter Amanda. This menagerie all lives in what was once the garden, and crap all over the patio.

Doreen begins final secretarial exams this week, and has a couple of interviews lined up at a building society and Montague's bank.

Gladys is already negotiating a percentage of the wage packet towards the housekeeping.

Wednesday May 13th

Cleaned m' wellies under the tap in the yard, and went off to a farm sale with Charlie. Charlie is interested in a trailer and an elevator, I'm after some sheep.

When we arrived Michael McMurdie had already begun to move along the line of machinery, — and the world and his wife were there.

Everything was 'on fire'. A corn drill petrified by last year's nitrogen, and a 'museum' thistle-cutter covered in hen muck, made almost new price. The bidding for a tractor that wouldn't start, — wouldn't stop!

Charlie got his elevator, — it had four flat tyres and will probably need a new engine, but he seemed to think it was a 'snip'. (Buyers tend to convince themselves of this, whereas the runner-up is invariably relieved.)

Cecil Potts got a trailer at an outrageous price, then went round telling everybody it was the only piece of equipment worth the money.

To the sheep ring. Michael made his usual introductory speech. 'Now gentlemen,' he says, 'these sheep haven't been abused in any way ... ' (a townie might be a little worried by the implication) ... 'they'll shift anywhere, all injected, ... ready to move ...,' and off he went.

Well, they weren't giving them away, that's for sure, — but I got a nice looking pen of twenty one-crop mules with twins early on, for about three quid more than I figured they'd make, — not too bad. I could hear Cecil telling somebody that the price was ridiculous, no money in them, only a fool would pay that for one-crop yowes. Then he went and paid the same price for some four-crops. I heard a phrase on the

radio once that fits Cecil, — 'he's got verbal diarrhoea and mental constipation'.

Charlie and I hung about to see a few cattle sold, but decided we'd be better off in the beer tent.

Didn't get home till after supper, a little tipsy. (Gladys wasn't impressed.) The sheep arrived after dark and the wagon driver just put them into the croft for the night.

Thursday May 14th

Checked the new purchases and discovered we had a lamb short. Did he get here? Was he loaded at the sale? Has somebody else got him? I should have checked at the sale, instead of gossiping in the beer tent, and pulling Cecil to pieces. I should know by now that whenever you're feeling pleased with yourself, watch out, — fate is getting ready to drop something on you.

Phoned the haulage, — 'oh yes, the driver counted them in, they were all there'.

Phoned McMurdie, — 'has anyone phoned in with an extra lamb from yesterday's sale?' 'No, not yet, — we'll let you know if somebody does. ... '

I count them again, twenty yowes, thirty-nine lambs. I walk around the croft, down the road, into the back field to see if he's crept through among the other sheep. He's not there, or here, or anywhere. I've been an incompetent fool. Serves me right!

At dinner time I'm still miserable, — just pickin' at the cold meat and taties when the phone rings, — it always rings in the middle of m' dinner.

'Morning,' says this unmistakable voice, 'y' know those sheep you bought yesterday, — hell of a price weren't they, ... well, I think I might have one of yours, — it's a lot smaller than the ones I got of course, — but I've definitely got an extra lamb, — it looks like one of yours ... red mark on its back. ... '

Cecil y' auld bugger,' says I, 'I'll be over this afternoon.'

Friday May 15th

Today is Rent Day. The day when simple peasants such as Charlie, Arthur, Percy and m'self duly pay our respects (and a big fat cheque) to the Colonel, our enlightened and compassionate landlord. The age-old procedure is that we lowly persons trot along to the 'Big hoose', clutching our forelocks, and hand over a few thousand quid, in return for a small glass of Cyprus sherry, and a licence to lose some more money for another year. Knickers is all charm and waffle of course (so would I be in his shoes) — his wife Clarissa glides about, serving drinks and asking us all how our wives are. She seldom listens to any reply, just smiles and glides on. You could answer that your wife had recently been savagely raped by a mob of Afghan extremists who happened to be passing by ... and the lady Clarissa would probably say, 'Oh I'm so glad, give her my kind regards, won't you. ... '

Lurking in the background is the Colonel's agent, Newcassel-Browne, one of the best lurkers I've ever seen. Of course he's the power behind the throne, he's the bloke who really decides if your hayshed roof is repaired, or your son gets onto the agreement. He's the Ayatollah.

We all left with a receipt, and notice that a Rent review would take place before the end of the year. Montague will be delirious about that!

Saturday May 16th

In spite of his pathetic performances so far, Willie is playing cricket again at some remote place in the hills, where the team are all shepherds and fast bowlers. They'll probably kill him.

Doreen is going to the YFC Spring Ball with Wayne, who comes to collect her in Cecil's Range Rover, tarted up in bow tie and evening suit. He's too tall for the trousers. His enormous feet and a lot of leg stick out at the bottom. He's obviously very aware of this, and keeps his hands in his pockets, pushing the trousers down on elasticated braces. 'Trouble is when he removes his hand, to take a drink or open a door or whatever, − the trousers shoot up again.

Doreen came down looking like a film star, dressed like a queen of the silver screen, a modern Marilyn Munro. There are bits of Doreen exposed that I didn't know existed (well I suppose I did really. ...)

Willie returns after dark, battered but triumphant. He scored eight not out and took two catches (obviously no pretty girls among the spectators). But the team lost, − all out for thirty.

Sunday May 17th

A week today we'll be in Corfu, − I've got Greek tummy already. Is there no escape??

Doreen lies in bed all day, Willie is limping, Gladys is singing 'Fly me to the Moon ... ' Sweep is missing, probably with the Pillick's Jack Russell or the Vicar's labrador. One of these Sundays the smiling congregation will come out of morning service, all uplifted, and there in the churchyard will be Sweep and one of his girlfriends half-way to paradise.

Took a walk around the farm after supper. It's a smashing night, warm and quiet. Lambs playing chasey along the hedges, cattle grazing contentedly, birds twittering, midgies dancin'. Not a man-made sound in the air. It smells delicious. Who needs a bleedin' Greek Island? England on a good night is unbeatable. There's maybe hardly plenty of them though.

Monday May 18th

Dosed the new ewes and lambs, and put them through the footbath, − without losing m' temper.

Rolled the hayfield in the afternoon, − stones and reaper blades are not compatible. It's a fairly boring job, but I like to see the straight pattern of light and dark green when it's finished, − looks neat and tidy, − as if somebody were at least trying to farm properly.

We have a lame bullock. Got him in, along with all the others in that field, dressed his foot and gave him a jab. I took the rest back, but kept old hopalong in the hemmel until tomorrow. He didn't think much of that, and bellowed all night. Another injection should do the necessary.

Tuesday May 19th

Not a good day. We have a yow with mastitis. It took me twenty minutes to catch the creature, by which time I was on the brink of a cardiac (again). Tied her legs together with a piece of baler twine, wrapped m' anorak around her head, and went home for penicillin and syringe. When I got back, the other yowes, presumably thinking my coat to be a bag of protein pellets, had dragged it off, and eaten half of it.

However, madam was still there. She was injected and marked for future reference.

It was nine o'clock before I got to the last field, and here I found another yow on her back in a rig bottom. She'd obviously been there all night, and was very wobbly. Another hour or so, and she'd have been a gonna I think. It took her quite a while to have a pee, and stagger off. Her lambs had little sympathy, — all they were interested in was breakfast, and they attacked the poor thing like a pair of piranhas.

Jabbed the lame bullock again, and took him back to his mates.

Wednesday May 20th

Returned to the muck. Promptly hit a pillar in the hemmel, and bent a prong on the fork. Am I fit to be in charge of any machinery?

Ordered a new water trough (and a new prong). This is called forward planning. I know from past experience that if I don't get these things fixed very soon, I'll be putting cattle inside in November, and find they have nowt to drink. Then next spring I'll come out full of enthusiasm to move more muck, and find a prongless fork.

Cecil has started silage. He puts it into those black plastic bags. It certainly looks simpler than haymaking between thunderstorms.

Thursday May 21st

Doreen has finished her exams. Claims she's failed miserably, but she has that in-bred peasant pessimism that always anticipates disaster, — so we'll wait and see.

Gladys is packed of course, labels on the cases, enough clothes to cope with a ten-man ascent of Everest (and it's 90° in the shade I understand). She has medical supplies to cure every disease known to science, − and a lot of foreign money, that looks absolutely worthless t' me. She's ready, on the blocks. I think if I threw her off the hayshed roof, she'd fly there herself no bother.

Doreen is locked in the bathroom trying on swim suits. Willie is banging on the door claiming he's 'desperate'.

That yow was on her back again this morning, the same one. She's determined. I'll be surprised if she's still with us when we get back. Charlie has promised to look after the stock next week, and take us to the airport on Saturday. He's a canny lad, Charlie, − we'll bring him a bottle of 'duty-free' back.

Friday May 22nd

Our last full day in civilisation, − perhaps our last full day *anywhere*. Will I ever see the farm again? If I do get back, will *it* still be here?

Sweep and I walk round the farm twice, fully expecting to find catastrophe around every corner, − maybe almost *hoping* for some major tragedy that would insist I stay in England for the next week.

But for once there are no sheep upside down, the mastitis yow is trotting about again (give her one more jab for luck). There's no foot and mouth, no fires, no flat tyres, no leaking water pipes. All I can find is a dead hen.

I suggest an outbreak of fowl pest, − but Gladys reminds me the hen was certainly fourteen years old, and probably very tired.

I think Sweep realises that this might be the last time he ever sees me, and follows t' heel all day, with big sad eyes watching my every move. Charlie will take him round the farm every morning with his own dog Moss, − and the two of them will show off like mad.

I'm not hungry, − just push m' dinner round the plate. I suggest I might be coming down with something nasty, −

but Gladys says it's just the excitement!

At night I walk the farm again looking for leather-jackets, disease and pestilence in the corn, anthrax in the cattle, bubonic plague in the sheep, pigeons on the rape, a hurricane on the horizon, . . . and end up in the Swan for a last drink.

Gordon promises to keep an eye on the garden, Jake, the postman, says he'll keep all the bills till I get back, Arthur will check for burglars, Walter Williamson the spray expert, say he'll inspect all the crops, and Percy says he'll shoot any stranger he sees up our road. They all tell me how lucky I am.

Back home Gladys, Doreen and Willie are drinking cocoa and watching the late film on telly. They are all dressed in tattered jeans, holey jumpers and no socks, — everything else is packed.

I take Sweep his last supper, and give his belly a good scratch, — he likes that.

Saturday May 23rd

After a restless night, up at 5.00 am. There's a blackbird on the hedge at the bottom of the garden tuning his beak and singing at the top of his voice. He's a proper show-off, but he can't half whistle. I've never appreciated him before, not really. However, this is it! I'm more or less resigned to whatever fate, Dan Air and Gladys may have in store for me.

Charlie took us all to the airport and abandoned me. Everybody in the building was going to Corfu, thousands of 'em, how can so may bodies get into one aeroplane?

Well I'll tell you how, — they assume every passenger is a five-stone dwarf with detachable legs! We were strapped in, then told how to get out in a emergency. The stewardess had to be jokin', — there was no way anybody could move, and long before we took off, all blood supply below the bum had ceased.

To my amazement (and relief) the thing got off the ground first try, — Willie keeping me informed of speed and height

(I didn't need to know). Doreen and Gladys anxiously holding hands, as if on their way to an audience with the Almighty.

We were offered drinks at 9.15 am! What did they know that we didn't? — was this the final snifter?

Some noisy tatooed Glaswegians in technicolour T-shirts didn't seem to care, — but then plummeting into the Alps may well be preferable to life in Sauchichall Street.

Later on they gave us a warm plastic box. If Gladys hadn't told me to open it and eat the contents, I wouldn't have known what it was. When I opened it I still didn't know what it was! Willie ate mine, most of Doreen's and Gladys's pudding while telling me that some French city was 35,000 feet below. (I didn't need to know that either.) I think I fell asleep, or perhaps the restricted blood flow rendered me unconscious, but the next thing I remembered was ear-ache, and Willie telling me we were landing.

By then the tatooed Glaswegians had mastered the art of sitting comfortably in an aeroplane, — they were legless!

Sunday May 24th

The Hotel Gigantos is a stack of concrete battery cages. Gladys and I are in cage 526. Doreen and Willie have their own cages somewhere else in the stack. We all have uninterrupted views into the side of a mountain, where another stack is being built. Each room has a balcony suspended over the room below. Gladys sits there dreamily gazing at the cement mixers, — but it doesn't look safe to me.

To the bar where the tatooed Glaswegians are in full flow, — or at least the ouzo and the retsina are in full flow. These are poisonous local drinks, specially formulated to render foreign visitors incapable of rational behaviour. I stick to gin 'n' tonic, and to be fair Costas (all Greeks are called Costas) gives his customers a very good measure (approximately half a pint).

Later, Gladys, dressed in her new floral costume, wallows

in the pool like a lethargic whale covered in chrysanthemums, — clearly visible I suspect from 35,000 feet.

Periodically she emerges from the water to lie on a sun bed, and read her copy of *Hollywood Husbands*.

Doreen is wearing three handkerchiefs and a smile of triumph.

Willie is missing. Apparently he was at a disco last night and has become allergic to bright sunlight.

Monday May 25th

The dining room is the size of an EEC intervention store, and here there is a garlic surplus, an olive mountain, and an oil lake.

Gladys who has a stomach fashioned of corrugated iron eats everything. She devours pink semolina (call Taramasalata),

enormous salads covered in goat cheese, bits of meat skewered onto knitting needles, octopus, swordfish, mince wrapped in rhododendrum leaves (she eats the leaves as well) sticky honey-coated cakes, green figs, and little spicey sausages that taste even more lethal than they look. She consumes ten times as much as she would at home, and there's no washing up. She talks incessantly, pointing out the honeymoon couple, the German and his fancy fraulein, the noisy Glaswegians. The Glaswegians and I have at least one thing in common, — we stick to steak and chips.

Doreen is into dramatic entrances, arriving late at the table in a more provocative outfit each night. She has acquired delusions of sophisticated maturity, only using her fork (little finger hoisted high) as she picks at a salad, or sips coyly at the dry white wine, with the occasional seductive dark-eyed glance towards a group of Manchester lads discussing mating tactics on a neighbouring table.

Willie's day begins at supper time, when he emerges looking as if he'd been in the lambing field for the last twenty-four hours.

Tuesday May 26th

Gladys is poorly, spends most of the morning in the loo reading *Hollywood Husbands*.

Doreen has gone off on a boat trip to a deserted beach with (among others) the sex maniacs from Manchester.

I go for a walk, but can find no grass, no sheep, no newspapers, no mart. I *do* find a bar full of Greek peasants. I feel we might have something in common, but they don't seem able to understand anything I say. I want to ask them what they're doing drinking and playing cards at eleven o'clock in the morning, — haven't they got a hemmel to muck out? I would've invited Willie to go with me, but he's still asleep, another rough night I suppose.

Wednesday May 27th

Doreen is bright red and very fragile. The three hankies have been replaced by a pair of white shorts, and a large loose T-shirt, with the undulating message 'Hands Off' emblazened across the front.

Gladys has decided to forgo any more figs.

At night I find Willie in the bar with his arm around a voluptuous purple-haired amazon (she's at least forty) and with what I assume to be her nubile daughter. I order a drink and consider confining him to his room, but when I turn around he's gone off with the daughter, and the amazon is eyeing *me* up. Gladys comes to the rescue.

Thursday May 28th   *Ascension*

I now know how the peasants can booze in the taverna all day. Greek agriculture is a whole lot simpler than ours, at least round here it is. They don't bother with disease-ridden winter barley or lame yowes or anything complicated like that, — here they have olives.

Olives grow on trees. The trees live for hundreds of years. All they've got to do is get the olives off the everlasting tree! Nothing simpler, they've got it mastered, fully automated. What they do (I saw them) is put a big net under the tree, and wait for a windy night, — if that doesn't get all the fruits off, they borrow an old Fergie from somewhere, and drive it into the tree trunk. This shakes the tree, and down comes the rest of the olives.

And then the master stroke. You might imagine (if you weren't familiar with Greek farming methods, that is) that the old peasant would now pick up the netfull of olives. Not on your life, — he sends his wife to do it! I've seen them to-day on the hill behind the hotel, gathering up the nets, loading them onto a donkey, and carting them off to the local olive press. Meanwhile the blokes are downing the ouzo in the pub.

Gladys is back at the pool again, and between wallows is discussing *Hollywood Husbands* with a Huddersfield wife.

Doreen is sitting in the shade with her hankies on, being chatted up by one of the Manchester lads.

The nubile daughter of purple amazon is looking for Willie, but Willie won't surface before dark.

Tried the Moussaka tonight, and a few bottles of red wine. Very nice.

Friday May 29th

Bad head today, — the Moussaka must've been off.

While Gladys was having a kip, I tried a quick dip in the pool (just so that I could tell Charlie I'd been in). Very few people about at this time of day, it's too hot, most are 'siesta-ing', so I can thrash about in my own inimitable style, and if I drown, then at least not too many will observe the fiasco. The tatooed Glaswegians are drinking heavily, clustered on an inflatable plastic armchair at the deep end. The purple Amazon sits like a contemplative Buddha in the shallows. I confine myself to a few spluttering breast strokes in the middle, careful never to be out of m' depth. My new bathing trunks (bought specially for this moment by Gladys) are less than satisfactory. They have a tendency to fill up with water every few minutes, and the weight then drags them down towards my knees. If I had some baler twine the problem could be remedied.

Then I see them, over on the rocks. The Manchester mob are there, and some other lads and lasses, and the lasses are all topless. Some are lying on their tummies and *I think* they're topless, others are on their backs and they're definitely topless. One of them is Doreen!

I hurry back to our room to tell Gladys to do something about it, but she's sitting on the bleedin' balcony, topless as well, reading the final chapter of *Hollywood Husbands*.

The quicker we get back to civilisation, the better!

Saturday May 30th

The last day, thank God. Tonight we'll be back in the real world.

Wake up early, set off for a stroll, and meet Willie backing out of cage 287 blowing kisses through the door.

For once we're all together for breakfast (we have to leave at ten). Gladys has packed, Willie is knackered, Doreen is exchanging addresses with a Mancunian. I pay the bill, and the bus comes at 10.15.

At the airport white people are disembarking, and brown ones are heading home. There is one exception to this otherwise accurate observation. Our Willie is still white!

The aeroplane is late. This gives Gladys a real chance to empty the Duty Free shop (presents for Boadicea, Florrie, Hilda, the postie, Tracey, Wayne, and a small box of chocolate liqueurs for Sweep). It allows time for Doreen to climb all over an unprotesting youth queuing for the Manchester flight. It allows Willie an extra hour's kip.

I find an English newspaper dated Thursday May 28th which looks exactly like the last one I saw before we left home. At least it confirms that England is still there.

It was half past five when I eventually limped onto home ground, on dead legs, and staggered into the terminal. Willie's case was the first one onto the carousel, the rest of our luggage was last. By the time we got to Charlie's car, our son was fast asleep on the back seat.

Sunday May 31st

Sweep seems very pleased to see me back, maybe Charlie's been a bit hard on 'im. He doesn't work very well for anybody else.

The yowes seem less enthusiastic, in fact positively apathetic. In my absence the one with mastitis has died (the lambs are in the croft with the tups). And the fox has devastated our poultry flock. Apparently Charlie shut them in one night unaware that Mr Fox was already in the hen house. When he opened the slide next morning out he comes licking his lips. Luckily for them, two old birds had been sleeping in the grain shed, and so escaped the massacre. Gladys says that a pair of superannuated hens is insufficient for our needs. She wants replacements immediately.

The cattle show distinct signs of improvement. Sometimes when you see them every day, progress is hard to find, but a week away and it's obvious. I don't expect they'll leave any profit, but they look canny.

It's great to walk on green grass again, I feel a new man. Gladys says it's the rest, the holiday, − but it's nowt the kind, − it's just being home.

The blood supply has returned to the legs, − m' bum is alive and well again.

Doreen and Willie sleep all day.

June

Monday June 1st

Walter Williamson had arranged to spray the rape today, —
I've forgotton why, — some killer bug I think. It's impossible to keep up with all these modern diseases and parasites,
or the antidotes. Anyway it's too windy, — they'll do it
tomorrow.

Florrie comes to help Gladys tidy up and gets her present
(a bottle of Greek brandy). She says she's usually a bit 'serup-
titious' about foreign drinks, but promises to persevere with it.

Willie goes back to school, — he's a wreck.

That yow's on her back again. Charlie says she never turned
over once while we were away. The quicker she's clipped the
better.

Jake brings all last week's post. No income, only bills and
technicolour rubbish. The latter goes into the Aga, the bills
can wait in the desk.

Tuesday June 2nd

Bought ten pullets from Arthur Thompson's wife for £1.50
each. They look no bigger than pigeons, but she says they'll
lay like good 'n's.

Doreen goes to the Poly to enquire about results, and is
told they should be out at the end of the week.

The sprayer comes to do the rape, — but it starts to rain,
they'll do it tomorrow. Walter says they should spray the wheat
for brown rust as well while the machine's here. Ye gods, is

there no end to this expense? I sometimes wonder what would happen if I sprayed nothing, — but I haven't the nerve to try it.

That yow's on her back again. I leave her there and go home for the hand shears. I disrobe the bitch. She looks quite embarrassed.

Wednesday June 3rd

The missing prong for the muck loader arrived while we were away, so I finished cleaning out the hemmel, and took three loads down to the village for Gordon, Davy Scott and Mrs Simpson. I considered tipping a load into the Pillicks' driveway, but I hadn't the nerve to do that either.

The sprayer gets to work on the rape and wheat.

I used to do this job myself with what would now be considered a 'toy' sprayer, a wee hundred-gallon thing with booms not much wider than my outstretched arms. You had to have four or five fills to do a twenty-acre field, — and it took half the day to fill up from the back kitchen tap. Then the pressure gauge went wonky, and I was always getting a nozzle caught up in the fence, and the safety officer said I was a serious health hazard. Now we get Walter's man with his special monster machine to do the job. His booms reach from the hedge to the middle of the field, — twice about and he's finished. (Well, almost.)

Thursday June 4th

Lambs scoured t' death in South field, several not 'doing' at all. Got them in and dosed the lot with the latest wonder worm drench, and beat them (and their bloody-minded mothers) through the footbath as well. As usual they either fly through, tip-toeing over the water, or they stand with their front feet in the trough, and refuse to move any further.

That foot-rot stuff always makes my eyes water, and by the end of the job they were red, and I was coughing and splutterin' like a calf with husk.

Gladys says she's not impressed with my aroma either.

Friday June 5th

Florrie wants more money (who doesn't?) Says she's on a mere 'subsidence' wage. She's obviously been watching some Scottish shop steward on the telly.

Gladys tells me that after lengthy negotiations over a lot of coffee and a lot of gossip, agreement was reached on a 20 per cent rise in pay, together with an extra hour's work per week. Both ladies seem to consider this a triumph. I reckon it has nowt t' do with me.

Saturday June 6th

Willie, the all-round athlete, played cricket today against Whittingdale, the league champions. They are league champions primarily because of their unique pitch, which is some way short of Test match standard.

Whittingdale play on Joe Anderson's old grass field. All home matches are preceded by the removal of the sheep who graze there, and then the removal of what the sheep have left behind. The pitch is then cut, and rolled and marked by Joe himself, who later acts as umpire. The fact that he is very short-sighted, wanders about in a permanent fog of Condor bar pipe smoke, and is unashamedly biased in favour of the home team, has always been considered irrelevant, — it's his field.

Our lot batted first and amassed thirty-five, twelve of which were extras. Willie batted at number nine (promoted up the order after his achievements of May 16th) and survived a series of bouncers, shooters and beamers for quarter of an hour, without actually putting bat to ball. Eventually he managed to hook a long hop to the square leg boundary, and got a thick outside edge over third man off successive balls. The next one pitched on line with third slip, and stopped dead in a wet patch. Willie went to kick it back to the bowler, and after a very confident appeal from cover point, was promptly given out LBW by Smokin' Joe.

You could tell Willie was disappointed.

After tea Whittingdale were soon in trouble, — their five best sloggers were back in the pavilion with only seventeen on the board, — and our lot in with the chance of a famous victory.

It was then that Smokin' Joe came to their rescue. Two blatant leg-befores were turned down with a sympathetic shake of the head, a clear catch to the wicket keeper was judged to have come off the batsman's pad, and when Willie took a skier at mid-on, Joe (a little late in the day) shouted 'no-ball'. Whittingdale won by five runs.

We all got back to the Swan by seven o'clock, and Elsie allowed Willie into the bar to have a pint with the team. She thinks Willie's lovely. What is it about that laddie?

Doreen is delirious. She's passed the secretarial exams, and can now look for a job. Wayne took her out for a celebratory nosh at an 'arm and leg' restaurant in town.

Sunday June 7th   *Pentecost*

Watched the farming programme.

There's this pontificating peasant from Pontefract standing in a field of thigh-high barley forecasting another record harvest. He's telling the whole world how many tons he's likely to get, how much nitrogen he uses, what he paid for the land, how many lambs he's sold, what his bank charges are . . . the man must be certifiable. These are secrets you wouldn't divulge to your wife, never mind to a million viewers! At any moment I expect to be told what his inside leg measurement is, and how many teeth he's got left!

I asked Arthur if he'd seen the bloke, and Arthur said he'd actually met him once, — played golf with him last year, and he was 'all wind and watta', — claimed he played off a five handicap. Arther filled 'im in five and four, and pocketed a quid.

The weather forecast says it'll be changeable ... with showers and sunny intervals. Obviously McCaskill isn't very sure what's coming.

Monday June 8th

Began 'cowin'' (cleaning) the yows.

This is no job for a sophisticated gentleman such as m'self. Some of the sheep are a disgusting mess, and by mid morning so am I, — muck to the elbows, wellies and glazey leggin's covered in dark green stuff, — two barrowloads of clarts.

Our sheep pens are right next to the road, so every now and then a passer-by can see me snipping and groping around the backside of some filthy yow. Jake the postman paused briefly to ask if I was happy in my work. The milkman lingered for a while, but left rather quickly when I lost m' temper, and threatened to cut the throat of a bitch that wouldn't stand still.

It was nearly dinner time when a very smart young lady, doing a survey on fertilisers, drove up in a spotless Fiesta. I saw her approaching out the corner of m' eye as I was bent over a particularly messy beast. She was dressed for a garden party, and carrying a clipboard questionnaire.

'Good morning,' she said, 'I wonder if you have time to answer a few simple ques...' but by then I'd straightened up and was firmly shaking her beautifully manicured hand, like any sophisticated gentleman would.

'Morning bonny lass,' says I, 'what do y' want t' know?'

She took one dismayed look at her mucky little mit, and said she'd call another day.

111

Tuesday June 9th

I had another visitor today. I didn't even see him drive up (the sneaky little git). There I was kicking some sheep through the footbath. I expect I was screaming some fairly colourful abuse at them, when this posh voice says, 'I believe we have an appointment for eleven. ... '

Jesus, I'd forgotten all about the VAT man!

'Oh,' I said (or something like that) 'I'd forgotten about you ... ' and I offered him my hand, — but he backed off.

I told him George Forster of Murphy, Forster and McNulty, did all the accounting, but he said he'd like to see a few invoices and receipts, and could we go into my office and look at the books.

The office? — well that creep on the farming programme last Sunday probably had an office, but I've just got an auld

desk tucked away in what used to be the dairy. In fact you can hardly find the desk, but it's in there somewhere, behind the cardboard boxes overflowing with bills and circulars, worm-drench bottles, old magazines, and a whole lot of other rural rubbish of one sort or another. There's a telephone as well, but you probably wouldn't find it if it didn't ring. Actually I know where everything is, provided nobody touches anything, — provided Gladys doesn't tidy up, — provided some nosey VAT Inspector doesn't start an 'archeological dig'.

I found a chair for 'im (and the desk). Gladys made him a cup of coffee, and we left him to it. At tea time he still hadn't come out, but Gladys thought she'd heard him sobbing. Then he emerged looking bewildered and bleary-eyed. He said he'd have to get in touch with Mr Forster, and come back some other day.

'Any time,' says I.

Wednesday June 10th

Typical English summer's day, — drizzling, and a bitterly cold wind blowing from the North Sea. Also a pretty cool telephone call from Montague regarding my alarming balance of payments deficit. Indeed our one-way cash flow is now so rapid, something dramatic will have to be done soon. We haven't sold anything since those geld yowes at the end of April.

Walter Williamson doesn't help the situation much by telling me some winter barley should be sprayed for mildew. Then the Water Authority produce a breathtaking bill which suggests we've got a leak somewhere, and there's a lamb in the back field obviously keen to be into the hereafter.

When I come in for m' dinner Gladys is listening to the radio, — Frank Sinatra is singing 'Life is just a bowl of cherries'.

Thursday June 11th

Doreen has a job!

She went to the Bank first, but when Montague learned who she was, he threatened to hold her hostage until her father paid off some of his overdraft. But Robbie the vet took her on as a receptionist secretary, and she starts there next Monday. Her job is to send out the bills, take messages, and generally help Robbie and Co. to cope with frantic farmers with a hemmel full of coughing calves, and townies with constipated pussy cats.

Anyway that's one dependant off the list, − no more pocket money, no more frock allowance, − in fact even a contribution to the housekeepin'!

Friday June 12th

All day dosing lambs, cleaning yowes, paring feet, and pushing everything through the footbath. By the end I was knackered and *smelly*. Soaked in the bath for half an hour.

To the pub after supper, smelling like a fairy. Cecil Potts, who is always first at everything has been clipping this week. Predictably he claimed there was a foot of 'rise' on the yowes, and the brothers Pringle (who do all the shearing in these parts) said they'd never clipped better sheep anywhere. (They say that to everybody, . . . except me.)

Went back to Charlie's place, and had a nightcap of his Greek brandy. I think that stuff would cure foot-rot.

When I got home at quarter to twelve Gladys was still up. She asked if I'd had a good night. She was all charm and smiles, and made me a cup of cocoa. There's a rabbit off here. I'm extremely suspicious of such behaviour. She's definitely got something up her sleeve!

Saturday June 13th

Gladys is still very chirpy this morning, singing 'Love is a many splendoured thing . . .' (at least I think that's what it was). At breakfast the bacon was really crispy, — perfect. Has she got a fancy man, I ask m'self.

When I came in for a cup of coffee after looking the stock she'd quietened down a bit. She was still smiling occasionally, but it had become somewhat forced by now, . . . did I detect a touch of uncertainty?

By the time Willie goes to play cricket, 'many spendoured thing . . .' has been replaced by 'Oh God our help in ages past. . . .'

At dinner time she's in the huff, total silence. A plate of mince and chips thumped down on the table, — half the chips in m' lap. The mince is cool and lumpy (like Gladys). 'Get your own tea,' she says.

So here's me wondering what's happened to upset the old

crow, when Doreen glides in (she's just got up) and gives me a little kiss on m' bald patch. 'Happy Anniversary,' she says.

What did she say? What day is it? Oh hell she's right, — it's our four hundredth wedding anniversary, or thereabouts. Don't panic, play it cool.

Later Doreen is in the bath, Gladys is feeding the new pullets (so far no eggs) and I phone a whispered reservation to the Gourmet. At tea time I say, 'So what's on the telly t'night then pet?' — No answer, just rattling crockery in the sink. I can't see her face, but I can picture it. 'Nowt I bet,' says I, ' . . . never is on a Saturday . . .' (pause) . . . 'Fancy a posh nosh at the Gourmet? . . . well it's a special occasion isn't it? . . . I bet y' thought I'd forgotten?'

We had a canny night, — but I must remember her birthday. When is it . . .? Some time in November I think, — make a note of it!

Sunday June 14th   *Trinity*

The brothers Pringle are clipping at Charlie's, so I go up to see how they're getting on. Seems they should be to our place some time tomorrow, — if it doesn't rain.

Doreen is preparing for her great leap into the working world tomorrow, deciding what to wear. Wayne phones, but she hasn't time to speak to him.

I get some sheep into the croft in case the Pringles come early tomorrow, and fix up the pens in the shed.

Those Pringle lads are something special. They're both as rough as heather, and have this ramshackle farm in the back of beyond somewhere, — only Jake the postman has ever seen it. He only goes about once a month and is met at the gate, never gets into the yard. He reckons the brothers are looked after by their toothless old widowed mother, who's never been off the place since her man died in '76. None of them can write their name on a bit of paper, and only ever deal in pound notes.

Still, they take no harm, always a good pick-up truck, and the best set of clippers money can buy.

Willie is at the Williamson's house all day, and doesn't get home till ten o'clock. He says he and Tracey just listened to records. ... All day?

Monday June 15th

Gladys takes Doreen into work. She'll get a lift or catch the bus after today, — but already there's talk of how a 'nice little second-hand Metro', would save everybody a lot of bother. (It never ends!)

Phone Hilda, who goes across the yard to ask how the clipping's going, and comes back to tell me the Pringles should be here after dinner.

Get some sheep inside ready for them, and another lot into the croft. It takes all bloody morning 'cos Sweep's in a funny mood. One minute he's biting and barking at everything, all at a hellova speed, with the sheep falling down thoroughly guffed, — and the next minute he's wandered off out of sight (dreaming of that randy labrador I suppose). I end up with a sore throat.

The Pringle brothers get here about twelve o'clock. They have their bait while they're hanging up the machines, cleaning them, sharpening and oiling. They never sit down. At 12.30 they're grabbin' the first victims, and they don't even pause for a fag till half past six. Finished! Tups, everything, — only one tit and half a lug missing out of a hundred and seventy-three ... not bad!

Luckily I had Gordon to help catch and wrap up, while Sweep and I ran back and forward taking naked sheep away, fetching fully clothed ones in. At the end of the day Gordon, Sweep, m'self and the sheep were dead beat. The Pringles gathered up their equipment, took their money without a word, and went to rattle off a few more at Arthur Thompson's before dark. I was never as fit as that, — never!

I remember m' father was a useful clipper. He would just stroll into the catching pen and grab the first sheep t' hand, no looking for bare bellies, — they were all the same t' him.

Once he'd got the yow cowped, he'd give it a right good hammerin' and more or less render it unconscious. Then while it was in a sort of coma he'd have the fleece off in three minutes flat.

Tuesday June 16th

Walter Williamson called to inspect some wheat. There's a lot of septoria about he says (there's always something nasty about according to him).

He tells me he and his wife have just had a weekend in the Lakes. 'Didn't get back till nearly midnight on Sunday, — had a grand time.'

'Didn't Tracey go with you?' I asked.

'No,' says he. 'She had a girlfriend over to stay while we were away. I expect they just sat about listening to records all the time, — y' know what these youngsters are like.' Indeed I do, — but I said nowt.

Tidied up the clipping shed. Gordon and I packed the wool and stitched up the sheets ready for collection.

Wednesday June 17th

Took the first of the grass bullocks to the mart, three of those left over after the wintering, including the alcoholic one. At last a sniff of income, the tide has turned, — a little. The cattle sold well enough, although the weights were a bit disappointing. Humphrey Smith, self-styled wheeler-dealer cum entrepreneur, bought the lot, and I was hardly out of the ring before he swooped down like a vulture, demanding luck money. He followed me to the office, niggling on about how he needed a tenner a head, stood over me while the mart secretary gave me some cash, snapped at m' heels as I came out into the alleyway, and when I counted out three fivers for him, he went berserk. Shouting and yelling, calling me some very uncomplimentary names, threatened never to buy any more cattle from me. It was all very embarrassing. After about five minutes of this nonsense I stuffed the money back into my pocket, and went to the canteen for a cup of tea.

Thursday June 18th

A rare glorious sunny day.

I get the reaper out and replace blades, oil and grease everything ready for cutting hay.

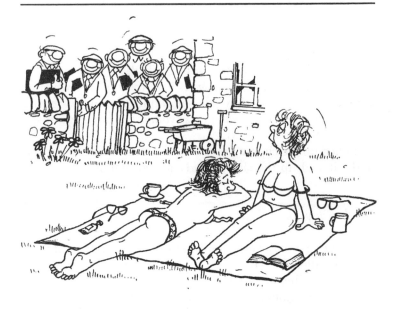

Walter Williamson's man is spraying mildewed barley, and reckons he should spray the turnips for yet another marauding beetle. Everybody else is, he says.

Gladys and Hilda are sunbathing in the garden.

A worm-drench rep (sorry, — animal health expert) who looked no older than Willie, came mid morning to advise me that all the lambs probably had some fatal disease, — and only *he* had the antidote. I gave him a similar message to the one Humphrey Smith got yesterday, and the youth left looking quite shocked, — but not before he'd paused long enough to get an eye-ful of the two lovelies lyin' on the lawn.

I can then only assume he went away and told all his colleagues within a twenty-mile radius the delights he'd seen, because by mid-afternoon, with the sun at its height, the yard was filled by goggle-eyed reps with their tongues hanging out.

At Gladys's request I sent them packing, and parked the muck spreader across the yard gate. I hadn't realised it, — but Hilda is a very tidy woman!

Friday June 19th

To the pub, — such a bonny night I walked there.

Davy Scott, Gordon and Mrs Simpson all paid cash for their loads of muck, so I bought them a drink.

Cecil Potts (genius, and father of Wayne) has twenty acres of perfect hay baled and led under cover. (So what's new?)

Saturday June 20th

Search among the turnips for beetles, — can't find any, but something is taking bites out of the leaves. Pigeons? They've actually been sitting on the bird-scarer, — it's covered in muck.

Phone Walter, — he says there's definitely a plague of beetles about, and it'll only cost a few quid to spray the devils to death, — better safe than sorry he says. (But he would, wouldn't he?)

Meanwhile West Indies are annihilating England at Lords. Willie and I watch the debacle on telly over dinner time. England's only hope would seem to be either torrential rain for three days, or the appointment of Smokin' Joe from Whittingdale as replacement for Dickie Bird.

Later in the day, Willie plays for the village, but as Brian Johnstone would say, he fails to trouble the scorer. . . .

Sunday June 21st   *Father's Day*

I get two very cheeky greetings cards from the kids, both depicting a very rich old man happily distributing goodies to his adoring offspring. I am reliably informed that Father's Day is exactly nine months ahead of Mother's Day!

All the signs are that the next few days will be fine. I base this bold conclusion on two things, — one, Wimbledon begins tomorrow, and I have to assume that the All England Lawn Tennis Association wouldn't buy all those strawberries if they thought it was going to pee down for a fortnight. And two, McCaskill is forecasting showers and bright periods. Consequently I decide to cut hay tomorrow.

Meanwhile I played golf with Arthur, against Jimmy Fletcher and his legal friend Rodney, — and I have to tell you we were rather lucky to win.

We were two down with three to play, thanks largely to an early purple patch from the solicitor, but on the sixteenth tee the match turned.

It was Arthur who did it. Placing his old grey ball on the peg, he clouted his drive with all the agricultural force he could muster. The ball never rose more than six inches from the ground as it sped to the ladies' marker about fifteen yards forward. It must have hit that object plumb centre, because it promptly came back like a bullet, and smashed into the men's tee box right next to Arthur (who was still poised statuesque at the top of his follow-through). From there the ball ricocheted, without apparently losing any velocity, onto Rodney's ankle bone, — and finally into a bed of nettles, several yards behind the point from which it had so recently begun its journey.

We lost the ball, but the opposition disintegrated, and we won on the last.

Monday June 22nd

Arose at 3.30 am to cut hay, determined to astonish the neighbourhood by completing the whole field long before the lazy slobs were out of bed.

As I should've known, this sort of ambitious nonsense only leads to disaster. I should've remembered the last time I went to mow a meadow in the middle of the night (it didn't work out then either).

At 4.00 am, on the second time around I hit an old pram and smashed a couple of blades off. It was a lovely still morning, the countryside was just turning over and stretching itself, deciding whether it could afford another ten minutes before it started the day, — but by the time I'd cursed, and hammered, and skinned m' knuckles, — the world was wide awake. Presumably the urban idiot who had dumped this bit of redundant domestic debris in our hayfield was still asleep somewhere. Percy would be for hangin' 'im, — and I'd be happy to pull the lever.

Gladys, aware that I must be having a few problems, brought some coffee out for me. She said nowt, — just left it at the gate and went home.

I struggled on and finished the field at tea time.

Tuesday June 23rd

Slept in!

This wouldn't have mattered too much if nobody else had known about it, — but Charlie came to borrow the reaper at eight o'clock, and I was still in bed. God, the shame of it, the humiliation. I'll not be able to look the man in the face for weeks (or until I catch *him* in bed some morning).

It's ridiculous, but we peasants are all the same, we hate to be 'caught' indoors during working hours. We try desperately to sustain the traditional myth that farmers toil from dawn to dusk and beyond, — calving cows, sowing corn, combining, shovelling muck, — anything. To be suspected of watching a Test Match in the middle of the afternoon is inexcusable. ('I was just doin' some paperwork. . . . ') Dozing too long after dinner is criminal ('I was just making a few phone calls. . . . ') To be discovered actually getting up as late as eight o'clock is an event likely to cause long and sinister laughter, and possibly even blackmail.

Charlie said nothing, but he was still grinning an evil grin when he drove out of the yard with the reaper.

Wednesday June 24th

Took some more cattle to the mart. Charlie said he was pleased to see I'd got there in time for the ballot.

They were a good trade, Humphrey Smith didn't buy any of mine, so he didn't get any luck money from me this week either. This is not necessarily good news however, — in a poor trade I might be pleased to see him bidding, — but for the time being he can get stuffed. It was also reassuring to find that the two butchers who *did* buy the bullocks were quite happy with the 'luck' I gave them.

Thursday June 25th

There are at least two schools of haymaking wisdom. One of them reckons you should scatter the newly cut grass all over the field immediately it's cut, so that it can die and become hay very quickly.

The other school figures that's all very well if you're hay-making on the Costa del Sol, but in the real world you could end up with well-wuffled muck very quickly. So this school allows one side of the swathe to lie for maybe a week till it's dry and crackling, before they turn it over to cook on the other side. Then it just takes a couple of days before you've got perfect hay. That's the theory, — in practice both systems can produce perfect hay or rubbish . . . it all depends on the weather, and lady luck.

My luck is out. Wimbledon is rained off. The Test Match is rained off. All haymaking is rained off!

Friday June 26th

Wimbledon is on again, Navratilova serving and volleying in glorious sunshine. The Test Match is on again, — Malcolm Marshall firing Exocets through a cloudless sky.

Back on the ranch however, bad light and a steady drizzle have stopped play for the day.

Saturday June 27th

The better weather has moved north, with a useful breeze to dry everything. The hay hasn't been touched, so no great harm done yet.

That's the good news, — the even better news is that Cecil Pott's perfect hay, baled and led a week ago, has fallen out of his hayshed overnight! This must be very disappointing for Cecil.

Sunday June 28th

Looked like another good haymaking day, so yoked up the acrobat, replaced a few tines, and waited anxiously for the dew to burn off.

Meanwhile Gladys and Doreen went to church (I asked them to put a word in for me). Willie has a job tidying up Mrs Simpson's garden. I hope he hasn't developed a fetish for older women. Knowing her reputation, I advised Willie to stay in the garden.

Turned the hay after dinner, and finished about tea time.

I've noticed I'm continually mentioning meal times (a wagon came in at breakfast time; went to the mart before dinner; down to the pub after supper). It sounds as if we do nowt but eat.

My memory tells me that certainly seemed to be the case in m' father's day. He would get up at six every morning and have a 'first breakfast' of tea and bread. He'd look the stock then have a proper breakfast at 8.30 (porridge and a massive fry-up). His dinner was always at 12.00 sharp (meat and milk puddin'). A quick kip and out again at one. Tea was at 3.30, sometimes taken to the field for him, and supper (another two-course feast) was at 8.00 exactly. He would have more tea and biscuits before he went to bed, — full!

Mother was either preparing a meal, or washing up after it. On a hay day or a threshing day she'd feed a dozen blokes. The Cambridge diet hadn't been invented.

Arthur Thompson and his wife Norma came over for supper. He says there's a rumour floating about that Humphrey Smith is knackered. (This is a technical term for financially embarrassed.)

Monday June 29th

At breakfast time a wagon comes for the wool, and Charlie phones to say he's got twenty yowes and lambs belonging to me in his hayfield. Sweep and I go to bring them back, and one of the yowes collapses half way down the road. She won't get up, so I have to leave her there. When I go back with the pickup to collect the old bitch, she's grazing on the roadside, and I have a hell of a job catching her. There I was kicking her into the pickup, telling her what I thought of her in simple peasant language, when the Reverend cycles past, all sweetness and light. ' . . . Morning,' he says chirpily, and quotes something about the one that went astray. (Does he mean the yow, or me?)

The hay situation looks good, another twirl with the turner and it should be almost fit to bale. After supper I phone Roger the contractor, and order the baler for tomorrow afternoon. Then go to the village in search of labour for stooking.

Tuesday June 30th

It's raining! — no comment. At least none that can be printed here, — but I think I can hear Charlie swearin' three fields away.

There'll be no baling for a day or two, so what can we do

instead? No good just wandering about in a bad fettle, cursing and kicking things. (But I do!)

In the afternoon brought the single lambs into the pens and weighed the best of them.

Got a dozen marked at twenty kilos or more, half-weight, — and entered them for tomorrow.

By dark, the rain has stopped.

July

Wednesday July 1st

A good bright morning. Lambs to the mart, — the first draw from this year's crop, they look good. Well . . . they look good till they get to the mart, then they just look ordinary among all the others. (Everything's relative).

Ghadaffi is in a right 'bolshie' mood, — knocking three kilos off everything, rejecting a lot of good sheep. Charlie has one turned down, but ours all get through at nineteen kilos. Not a great trade, but the subsidy (when it comes) should make them a decent price, eventually.

Rush home and wuffle like a mad thing. The wuffler is past her best, and disintegrates at four-acre intervals, but survives the ordeal, — just.

One of the new pullets has produced a marble!

Thursday July 2nd

The twelve yowes who were deprived of their offspring yester-day are in the garden at three o'clock this morning, bleating pathetically. When I stick my head out of the bedroom win-dow to suggest they might prefer to bleat elsewhere, I discover it's raining.

Gladys, needless to say, sleeps through this little drama, mouth open, making almost as much noise as the sheep.

Sweep and I put the yowes into the hemmel, then we both go back to sleep on the sofa until breakfast. By the time the rest of the family are up, so is the sun, and by afternoon I can wuffle again. (This haymaking job is like *Eastenders*, — it can go on for ever without really getting anywhere.)

Friday July 3rd

What we need is sunshine, — but it's cloudy, and not enough wind. I move the hay again, and at tea time we have a go with the baler.

After three bales we are convinced that the hay is not quite 'fit' enough. This conclusion is arrived at after three snapped sheer bolts, and a bad back while trying to lift one of the bales. Only the strings move, the bale stays put.

At the pub Arther tells me he's had no rain last night, and has managed to get ten acres of really good hay. Charlie hasn't touched his stuff yet. Cecil Potts has put his superb hay back into the shed (with a lot of props to keep it there). Percy only makes silage, and says anybody who makes hay in this climate should be certified. He could be right.

Saturday July 4th   Independence Day, USA

This looks better, — the sun is shining, the breeze is blowing, and it's all systems go. Willie will not play cricket today, — he will windrow the hay in front of the baler. Gordon and I will stook bales.

Charlie has some fancy equipment that stooks them mechanically. Jimmy Fletcher swears by those big round bales. But we're a bit auld-fashioned I suppose, and do it the hard way. I still reckon part-time labour is cheaper than part-time machinery, and maybe it doesn't depreciate quite so quickly either.

Surprisingly all goes well. There are a few burst bales to put through again at the finish, the dyke back ones are a bit heavy, — but on the whole it's a fair crop of canny hay. We had it all stooked by dark, and went down to the pub for a pint.

Anybody who hasn't got their hay baled tonight is a long way behind.

Sunday July 5th

It's a fact, just when you think you've got things under control, when everything seems to be going right, when farming looks easy ... beware! — just around the corner there's a cynical little gremlin waiting to kick you in the goolies.

Today he's definitely after me. ...

(a)   it's chuckin' it down, torrential, — it'll go straight through the bales.

(b)   the yow that collapsed on the way back from Charlie's has collapsed again (permanently).

(c)   there's a bullock running up and down the wrong side of the fence, — trampling acres of winter barley.

(d)   Walter Williamson discovers another wheat disease called fusarium.

(e)   Gladys's mother comes for her dinner.

I read in the Sunday papers that there's an alarming increase of suicides among farmers ... (really?)

Monday July 6th

There's a gale blowing. Took the top bales off the stooks and set them up to dry out, — they'll all be bent like bananas tomorrow.

Sprayed nettles and docks around the farm, with the knapsack on m' back. Florrie says I'm killing all the rare plants and the 'conversationalists' will be upset. She watches David Bellamy too much.

A letter from Newcassel-Browne informing us that he will call next week to discuss the rent. (We must be prepared for this visitation.)

Tuesday July 7th

Got some ewes and lambs into the croft for the mart tomorrow, together with a superannuated tup who has chatted up his last mule yow. Poor thing, another night of passion would be too much for him.

Went to get Charlie's elevator (we share these sophisticated items whenever possible) put some straw on the bottom of the hayshed, attached the bale carrier to the tractor, and filled her up with diesel. All this chief needs now is a few Indians, and we're ready to lead hay.

After tea Gordon came up, and Willie arrived home from school, so we made a start and got about a thousand bales in. We'll hopefully do the same tomorrow night, though we might swop jobs 'cos Willie was getting a bit weary at the bottom of the elevator. I'll give him the tractor tomorrow, and bribe him with a man's wage.

Wednesday July 8th

One of the lambs in the croft, drawn out ready for the mart, a prime animal, a superb beast in every way, as fit as a lintie yesterday, — is lying deceased! Stiff!

I am very upset, and jump up and down on the carcass, screaming every obscenity I can remember from an extensive rural vocabulary. This confirms Gladys's opinion that she married a maniac, — but fails to revive the lamb!

Of the remainder, one is rejected. How can he reject a good healthy lamb of twenty kilos taken straight from his mother? The tup staggers through this last ordeal, and raises ten quidsworth of sympathy.

We lead more bales at night. Willie drives like a mad thing, but only cowps two stooks.

Thursday July 9th

The forecast is lousy, — so I spend all day leading the rest
of the stooks home, and park them in the grain shed and the
hemmel, under cover. We can finish stacking any fine night,
and if Charlie wants his elevator back (he was baling yesterday) at least our stuff is safe.

It's not the best hay we've ever made, but it's not bad. A
lot letter than the black muck we couldn't even burn last year.
Maybe not as good as Cecil's, — but at least it's staying where
we put it . . . so far.

Friday July 10th

One 'eye' in the hayshed has developed a bulge overnight,
a kind of instant pregnancy. I cannot contemplate the shame,
the humiliation, of my hay falling out of the shed, so I spend
most of the morning installing a forest of props.

Took a walk round the corn fields after dinner, and found Walter groping about looking for galloping root-rot, or some other tragedy he can spray away. He has a bunch of wild oats in his hand. There's another stimulating job coming up soon.

At night Gordon, Willie and I stack the last of the bales, — very carefully, — more props and a lot of baler twine. How did we keep the farm together before we had baler twine?

We go down to the Swan to celebrate. Willie is allowed in for a half-shandy, then feeling like a fully fledged peasant (and with a few quid in his pocket) he goes off to impress Tracey.

Saturday July 11th

After being rewarded for his haymaking efforts, Willie has now developed a taste for money, and has offered his services as a wild oat picker at weekends and after school, — provided I show him what a real wild oat looks like.

Today he's playing cricket again. The team must be desperate, he's hardly shone so far, but he says he'll throw caution to the wind this afternoon, and slog his way out of the bad patch.

The other great sporting event is the Church Garden Party. Doreen is in charge of the fruit and veg stall, dressed as Nell Gwynne. Gladys runs the jumble department (dressed as Gladys). It's all in aid of the Roof Fund.

I avoid this fête, and spend some time preparing my defences against the imminent attack of Newcassel-Browne.

Cecil Potts sent a pig to the fête for visitors to guess its weight, but it escaped, and galloped squealing through the vicarage garden, ate the cake stall, and dug up the Pillicks's lawn. The policeman on car park duty eventually caught it as it was about to dig up the patio as well.

Willie was caught at deep square leg for eighteen, — scored off four balls!

Gladys came home with three pairs of psychedelic wellie socks for me, and Doreen returned with a bottle of whisky,

having correctly guessed the weight of the runaway pig. She denies it of course, but there is some suspicion that Wayne may have leaked the vital information.

Sunday July 12th

Moved all the decent, respectable machinery down into the wood again (where we hid it from the Safety Inspector) and instructed Gladys to remove all signs of prosperity (e.g. TV, washing machine, fridge etc.) prior to the Ayatollah's visit tomorrow. I want him to have the impression that he's damned lucky to be getting the existing rent, never mind a rise.

Willie and Tracey are picking wild oats (sometimes).

Monday July 13th

I feel Gladys has gone too far.

She is dressed like a refugee and is doing the washing out in the yard using the old mangle she's unearthed from somewhere. The electric kettle has vanished, and we can't offer Newcassel-Browne a cup of coffee because (she says) we never light the fire until November. She's switched the lights off at the mains as well. I suggest a small whisky, but all I find is half a bottle of turnip wine, given to us by Jake the postman in 1958. The sofa has gone. Sweep is eating a bone in the passage although normally she never allows him into the house.

All this presents a pretty desolate picture, but Newcassel-Browne is unmoved, — he just twitters on about how refreshing it is to find a household which still preserves the fundamental values of a by-gone age, — how reassuring to discover a family who can still survive without a dose of Wogan every night, — how delightful to come across simple sound economics in a plastic world. . . .

I have no idea what he's on about, but I can tell our plan hasn't worked.

He demands a ten per cent increase. I insist on a similar decrease, and after much discussion, argument and more turnip wine, — we compromise.

From next May the rent goes up ten per cent.

Tuesday July 14th

Inspect the winter barley for wild oats, and discover that Willie and Tracey have missed most of them. This is explained by a small area of flattened barley, where Willie and Tracey have obviously 'rested' a while.

They're sacked, I'll do it m'self.

Wednesday July 15th

Took lambs and cattle to the mart.

The lambs all got graded and were a canny trade. However, cattle prices were well down. There were just too many for the butcher's mafia, — they got filled up, then stood about laughing at each other's jokes. Anybody late on the ballot was unlucky. I was unlucky and brought two home. (It's the old dilemma. What do you do? I can't afford to give 'em away, — but I can't afford *not* to sell them either!) One of mine never got a bid. He began coughing as soon as he entered the ring. I swear he was smiling when he got back home.

Back to the wild oats in a high wind. After three hours I'm going blind, now you see them, now you don't, — the bloody weeds won't stand still long enough.

I pick imaginary wild oats all night, disturbing Gladys. We're both knackered by morning, and decide to get up at five o'clock.

Thursday July 16th

Top-dress the hay ground, — we'll put the lambs onto this when they're spaned. Well, that's where we'll put them, — whether they'll stay there is another matter.

Fill in the dipping form and send it off. I've told the little bureaucratic twits that we intend to start the operation at 5.00 am on Sunday morning, — surely this will discourage anybody from turning up with a stopwatch.

Pull more wild oats. It begins to rain while I'm at the far end of the field, and I'm soaked to the Y-fronts by the time I get home.

Friday July 17th

An embarrassing episode.

Whenever I'm working by m'self, I tend to blather away to m'self (doesn't everybody?). I might have some imaginary negotiations with the Ayatollah, a desperate debt-ridden discussion with Montague, or a straightforward verbal punch-up with Gladys. Sometimes I even speak for both m'self and whoever the opposition might be. It can become quite heated, but of course I always win the argument.

Anyway, there I was stuffing wild oats into a plastic bag and rabbiting on to some invisible companion, when a voice from over the hedge says, 'what a load of rubbish,' — and there's Charlie grinning all over his face. It's not the first time he's caught me. I remember I was looking at some cattle one fine summer night in the field next to his place. We had this canny quiet bullock that always came up to me, and I used to scratch his lugs and pass the time of day with 'im. On this occasion I put my hand over the bullock's eyes and said, 'whoops, it's gone dark already, — the nights are cuttin' in quickly aren't they. . . . ' (it was a joke). Then I heard the giggling behind me, and there's Charlie sitting on the gate.

He told Gladys she should have me certified, but she said

I was too far gone, and there was nowt anybody could do.

To the Swan after supper, where I took 50p from Charlie at dominoes, and felt a little better.

Saturday July 18th

Dipping!

Gordon comes to give me a hand. We get all the sheep in, and start about ten o'clock. We draw out twenty lambs for Wednesday's mart first, and they escape the bath.

We haven't got one of those clever new-fangled baths, where the unsuspecting sheep have little option but to dive in. Ours is the old-fashioned variety, with a hole by the side where I stand, wellies full of water, and turn the sheep over as Gordon reverses them into the bath. By the finish, my eyes are stinging, and Gordon is a wreck.

The lambs are little bother. Gordon just holds them over the water and drops them in. They go straight under, and I fish them out and point the panic-stricken things towards the dripping pen.

The yowes are a bigger problem. They've seen it all before, they can smell what's coming, they dig their toes in and fight like hell to avoid the watery drop. Of course you always get the odd one who figures she can leap clean over the tank, and only realises in mid flight that she ain't gonna make it, — and then the man in the hole gets half drowned.

The tups are the worst. Far too strong for Gordon, they just stand like concrete monuments. He'll be pushing at one end, and me pulling a back leg at the other, until we get the stubborn sod over the edge, — to the point of no return. Then he's had it, but he's got to go over first time. And to hell with the rule that every animal should be in for a full minute ... some of our auld yowes could never survive that. In fact several mature ladies have to be dragged onto dry land, — the weight of the water in the fleece is too much for them, — they can't even shake themselves without falling down!

We finish the job at one o'clock, by which time Sweep is missing, — but I know where he'll be, and sure enough he's hiding in the hayshed. He whined a lot, and looked pathetic, but he went in. It'll do him no harm, — keep the flies off him for a while.

Sunday July 19th

Woken at 5.00 am by a bespectacled bureaucrat with a notebook and stopwatch. It's raining, and he's standing dripping at the back door.

'Well,' says I, 'y' see we knew it was going to rain today, so Gordon and I did them yesterday. ...'

I gave him all that stuff about getting casual labour at weekends, and yes, of course we'd given them a full minute. I showed him the empty dip can, and gave him a cup of tea.

Monday July 20th

Heavy, warm, cloudy day, — thunder threatening. A good day for lambing time maybe, a good day for growing grass, — but not much use if you're still trying to make hay, like Davy Scott. He's always last. Doesn't seem to make much difference though, — he always gets there in the end, never worries. He'll probably live a lot longer than the rest of us.

Even Florrie is lethargic. She blames 'the humility', and sits about drinking tea and gossiping with Gladys all day, while Sweep and I wander up and down the tramlines, me talking to m'self and pulling up the occasional wild oat.

Of course one of the dangers of talking to yourself is that you're sometimes in another world. I was just having a little rest at the end of the field, daydreamin' I suppose, not a sound ... when two bleedin' jets came over and nearly took m' cap off. I think Sweep heard them before me, 'cos he ducked, but he never said a word, — and I nearly had a heart attack!

Tuesday July 21st

Walter Williamson's man sprayed a field of barley to kill the wickens. It's not too *bad* really, seen worse, but it's right next to the road, and will certainly be an embarrassment by next year if we do nowt.

Walter and family are going on holiday tomorrow. (They're taking Tracey with them this time, which is very wise, — she'd organise an orgy every night if left at home on her own.) Anyway he had a good check round all the corn, and for once failed to come up with any convincing arguments why I should spray anything.

This is a considerable relief, — or perhaps I'm just beginning to resist some of the 'expert' counsel thrown at me from all quarters. When we were younger we were assured that we would become wiser as we got older, and I think we accepted that, more or less. We swallowed whatever the older generation gave us, castor oil and criticism. Now I'm not convinced that age and wisdom necessarily go together, — at best a kind of arthritic canniness comes over you. But in farming a lot of science and engineering are beyond my comprehension, — how on earth does a simple working peasant keep in touch with computerised combines, a bewildering collection of killer chemicals and a metricated mountain of medicines for mule yowes? Y' have to depend on some 'alleged expert', some of the time.

Wednesday July 22nd

Drew out the undipped lambs for the mart, and decided to spane the rest.

Humphrey Smith, the wheeler-dealer and enthusiastic collector of luck money, is not in the ring today. Rumour has it he owes the mart a lot of money, and has gone into hiding somewhere beyond Tow Law. (I didn't think there *was* anything beyond Tow Law.)

The lambs made 4p a kilo less than the national average for last week, and two were rejected. Arthur Thompson has about a dozen rejected every week. He just leaves them on the old railway line behind the mart, and presents them to Ghadaffi a week later. Sometimes it works, sometimes it doesn't. He currently has fifty running about on the embankment, eating cinders and weeds in the hope that they'll become thin enough to grade.

Back home I dosed the remaining lambs and took them over to the hay field. Secured the gate with barbed wire, piled some old troughs along the bottom, and left their mothers in the hemmel for the night.

By six o'clock the racket had begun in earnest, — the yowes in full voice calling for their offspring, and the lambs standing at the hayfield gate, bleating back. We'll never get to sleep tonight.

Thursday July 23rd

The yowes are still making a terrible row, primarily because six lambs have found their way home, and are trying to get into the hemmel. I briefly consider shooting the little buggers, but instead throw them into the pick-up and take them back to their mates. For a while I couldn't find where they'd escaped. The gate was still secure, and the fence is all pig netting down this side. Then I saw one of the six wanderers go straight to the water-trough. He was about to climb onto the ball-cock chamber and jump over the fence, but I just beat him to it, and felled him with m' stick. Two rails required, — and a new stick.

Charlie has been injured by a suckler calf. Apparently he had it in the crush to dress an infected foot, and it kicked him in a delicate area not far south of his belly button. Hilda says he's alright, but it took his breath away and made his eyes water for an hour.

Friday July 24th

Let the yowes out, — they're ravenous now, so they will hopefully settle down and just eat.

Went to the Swan after supper. Cecil Potts has cut a twenty-acre field of winter barley. Says it's eight tonnes to the hectare (whatever that means) all at 14 per cent and sold for malting already. That's just like him of course, every egg a double-yolker.

Charlie reckons the crop couldn't be ripe, it must've died of some rotting disease. Charlie is still a little tender after his accident, and says from now on any lame calves can just limp.

Willie breaks up for the summer holidays. We await his school report with interest.

Saturday July 25th

We have a yow looking very depressed. I quickly deduce that
this is not the result of frustrated motherhood, but rather one
of the many obscure and incurable afflictions peculiar to sheep.
The symptoms are as follows, — head down, lugs droopy,
slavering, glazed look in the eyes, wobbly gait. However, when
I try to catch the creature to dose it with Robbie's patent elixir,
it bolts off leaving me clutching a handful of wool, and runs
at great speed into a tree. This is not a happy sheep.

Willie plays cricket against Belham. Nobody likes playing
them, because they have a psychopathic fast bowler called
Freddie Johnson. Freddie is forty-seven and circular, but
generates terrifying pace and venom from a thunderous run
up, which begins by the sightscreen. The trouble is no one
is ever very sure where the ball will go, and on these rural
pitches anything can happen. A yorker can leap up and hit
a disappointed batsman in the teeth. The next delivery, pitched
just in front of Freddie's left foot, might stay low and break
the batsman's toes. And that's if the delivery is straight. It
has been known for mid-off to take evasive action.

When Willie went in our lot were twelve for eight, and three
of the eight wouldn't be fielding. He reckoned the earth trem-
bled as Freddie approached the stumps (and so did Willie I
expect). The first ball cleared the wicket-keeper and beat the

long stop as well. The second was a wide, but it hit one of the slips and he had to have treatment. The third got Willie on the toe and he was out LBW, in spite of the fact he had fled half-way to the square leg umpire.

We lost.

Sunday July 26th

A quiet day, — the lull before the storm perhaps? It'll be harvest soon (the bold Cecil has almost finished his winter barley) and I always get a bit anxious at this time of year. At the mart and in the pub they're all talking about bumper crops, and how much it is per ton, and generally counting up the money before they've even greased the combine. It's worth nowt till you've got it in the shed!

Willie is away on his old battered bike somewhere. He never rides his bike. What's got into him? Is it yesterday's blow on the toe? Is the poor lad lame?

No, I should've guessed, — Doreen it is who reveals that as his beloved Tracey is away on her hols, Willie has transferred his fickle affections, temporarily perhaps, to a voluptuous fifth-former called Dawn. Dawn, we are told, is well known for her interest in wild life, and is regularly accompanied into the countryside of a Sunday by any one of a dozen enthusiastic spotty biologists to observe nature, and Dawn maybe, in all her splendours.

Willie's fascination with the flora and fauna may be short-lived, — he comes home knackered, carrying a sick bike.

Monday July 27th

The depressed yow is no longer depressed, — and the lad from the kennels picks her up after breakfast. It is some small consolation to see he already has five bodies aboard.

One of our lambs has tried to hang himself in the wire netting, — obviously been there all night with his head through into the wheat. As I approach the fool goes crackers, pulls the netting off the posts, and runs away wrapping himself up in the stuff as he goes. He doesn't get too far, and by the time I have him disentangled I may have persuaded him it was a silly idea in the first place, — perhaps.

Cecil is swathing his oilseed rape.

Tuesday July 28th

Got all the yowes into the pens, and selected those who will not have the privilege of my company next year. Examine dentures and undercarriages, and find four who will certainly be unable to suckle a lamb in future. Several others are entirely toothless, and as for those who have just retained the odd insecure incisor, — I suspect a good sneeze would remove *them*.

Of the total number to be cast out, a couple are seriously anorexic, one has a belly rim trailing on the ground, another has no ears (removed no doubt during a frantic clipping day by a Pringle) and yet another trembles alarmingly while walking on her knees.

We'll need about forty new young breeding sheep.

Wednesday July 29th

To the mart with this unique miscellany of geriatrics. The quite respectable ones make a quite respectable price, the poorer ones make a canny price, the rubbish make a bob or two, — and the knee-trembler gets a lot of laughs. Arthur Thompson pretends to blow a fanfare as she enters the ring, and a cynical butcher shouts olé.

The whole lot average fifteen quid, — I'll settle for that.

Thursday July 30th

Gladys shows me Willie's school report. She's been hiding it for a week, and when I see it, I can understand why. Words such as commitment, discipline, concentration and application occur frequently, as being qualities he's apparently lacking. His contribution in the classroom is described as rebellious, bordering on anarchic. This term he is credited with seventeen black marks for disorderly conduct, ten of them after being apprehended in the girl's changing rooms.

Ah well, nobody's perfect, — but it seems prospects for the GCSE are gloomy, and our early hopes that he might end up as a solicitor, an accountant or even a government grader (the ultimate dream perhaps) are fading fast. It's pretty obvious he's going to be a peasant, — if he's not locked up first!

Friday July 31st

I can hear combines humming all around. Can't actually see any, but I assume every engine noise is a neighbour starting his harvest.

I phone Charlie at dinner time to ask him how long before *he* starts. Hilda says he's gone to the mart! The sneaky sod, he never goes to the mart on his own. What's he after? Should I be there too? Is something going cheap?

Walked round all the winter barley, rubbing out samples, and chewing on the pickle until I was sick of the taste. The field that we sprayed is nearly ready I think. Discover I've left the batteries in the moisture meter since last October, and the whole thing's corroded t' death.

Took a few handfuls across to Arthur's, and got it tested on his fancy electronic gadget. It says 20 per cent!

Went to the Swan after supper to get all the local rural wisdom. The brothers Pringle go to different pubs, — Alfie comes to the Swan, and Geordie bikes over to the Crown.

Then they go home and compare notes. They know everything about everybody for miles around!

Charlie says he's bought a wagon of charolais stirks for nowt, — nobody was there, he claims, — all combining.

Cecil Potts certainly was. He's finished his winter barley (all at 14 per cent and just over three tons, naturally).

Arthur says he might cut a trailer load tomorrow, just to see what it's like. Charlie reckons he'll be cutting by Monday. Even Davy Scott is talking boldly of starting next week, and nobody starts *after* Davy.

This is serious. There's nothing more likely to create stress in a peasant than seeing all his neighbours combining away on a fine day, while he's got nothing better to do that cut the lawn.

August

Saturday August 1st

A glorious summer day.

Had another look at the barley. Took another sample to
Arthur, — this time the magic box says 18 per cent. Took
another sample from the same field to Jimmy Fletcher's: 22
per cent. Took a third to Northern Grain and got a reading
of 20 per cent. What do you believe?

In the afternoon went for a ride around the countryside in the pickup to see what everybody else was doing. I remembered as a little laddie being slowly driven along country lanes of an evening, with m' father looking intently out of the car window at some other farmer's fields. Although I didn't realise it at the time, those trips were an early introduction to an essential part of farming psychology, — that is, how's the other fella doin'?

For instance, when things are going badly, the fractured ego of the depressed peasant can quickly be repaired if he finds the bloke down the road has a dead yow as well, or a poor crop of barley, or a mouldy stack of hay steaming in full view of the traffic.

On this occasion I was distressed to find several combines going flat out in clouds of dust, — all very worrying. Went home and had another chew at the barley.

Sunday August 2nd

Another fine day. That's four on the trot! Phoned Roger the contractor, — he's combining at Fletcher's.

Couldn't eat m' roast beef and Yorkshire puddin' for worrying. I feel I should be cutting corn. Everybody else is, — mine *must* be ripe.

The rest of the family go off in the car to see Gladys's mother, and I'm left at home, biting m' fingernails and more barley. I'll smash the false teeth before this is over.

In the end I can't stand the tension any longer and drive over to Jimmy's. They've finished a twenty-acre field. It looks no riper than mine. Roger says he can be at our place by Tuesday. Right, we'll have a go then.

Gladys and Doreen to the church in the evening. I tell them to pray hellish hard for a fine week. Willie promises to put in a good word for me as well, but I'm not convinced *he'll* have much influence.

Monday August 3rd

Final preparations in the grain shed. Checked trailers and augers. Grease is bubbling out of every bearing, the tanks are full of diesel; everything is rarin' to go.

Walter Williamson, bronzed and relaxed from his holiday in Spain strolls in to tell me his man will desiccate the rape this afternoon, and asks me why we haven't started the barley yet. 'Looks rotten ripe,' he says.

Florrie is eager to tell me she's seen ten combines going today, and hangs the washing out in a warm breeze.

Phone Roger again, late at night. 'Yes,' he says, 'in tomorrow, about dinner time ... (if he's not, I'll hijack 'im!)

Tuesday August 4th

The forecast is not encouraging.

I pace about all morning waiting for the sound of Roger's approach.

Hurry round the stock in case he arrives early (some hope). There's a lame bullock, but he'll have to wait for treatment.

By dinner time the clouds are rolling in from the West, — but Roger isn't. At four o'clock I can't stand it, and go off in the pickup to look for him. They've just finished and are sitting about drinking tea and telling jokes. It's a difficult situation. My first inclination is to go home for the twelve-bore and shoot the lot of them, — but I pretend to be very calm.

'Be with you in half an hour,' says Roger. 'You weren't getting worried were you?'

'No, no,' says I, as cool as can be. . . . 'Just passing this way, thought I'd look in and see how you were gettin' on. . . . '

By 5.30 he has the table linked up, and off we go. It seems to be running nicely. I get a full trailer load after the first time around, — it feels hard, a nice sample. Then the spots hit the windscreen. At 6.30 it's bucketing down!

Wednesday August 5th

I am not in a good mood.

I take out m' frustration on Sweep, the lame bullock, Jake the postman, and any other innocent that happens by. Gladys, who's seen it all before, says nowt.

The combine stands idle, shiny yellow in a grey drizzle.

You feel obliged to do something useful in such circumstances, so a bit late in the day, perhaps, I decide to draw out a few lambs for the mart. While I'm getting them into the pens, one of the tups crashes through the fence and begins to ravish everything. By the time I get him out, we've missed the ballot at the mart, and get a moderate trade at the end of the sale.

Everybody is talking about record yields and low moisture content. Cecil has sold all his for malting straight off the field. I feel distinctly inadequate.

Thursday August 6th

A better day. The rain has stopped, but we need a breeze. Roger is getting anxious because he's supposed to be somewhere else by now. This means we'll start cutting again before ours is properly dry, — I won't be able to hold him back, or he'll pull out.

At supper time Doreen tells us how boring her job is. I remind her that she gets paid, regardless of the weather. Any ideas she may have about early retirement for a more exciting life on the dole will be viewed with little sympathy in this house. She goes to bed in the huff.

Friday August 7th

The sun is shining, the breeze is blowing, Roger is whistling, and the combine moves off at eleven o'clock in perfect conditions, followed by Gordon who has come to help me lead off. Roger is going like a train, Gordon hurries back and forth to the field. It's all go.

Then disaster! To save time I decide to take over from Gordon for a couple of loads while he gets his bait. (Roger always eats his on the move.) As I've said before, I'm seldom entirely happy with any sort of mechanical job, but we can't afford to stop for something as unimportant as food. Other people have this simple task off to a fine art. They just drive up alongside the moving combine, get into the right place, and the right gear, collect a tankful or two and drive off again, — no bother.

But for me it's all very difficult. I've no confidence. I'm too close to the combine or too far away, I'm going too fast or too slow, I look back to check the trailer and drift off line, I concentrate on the line, and find the grain belting down m' neck. This time I was determined to do it properly. Even young Willie does it perfectly. Ye gods I think Gladys managed a load once, — surely the boss can do it.

So there we were, going as straight as a die, absolutely parallel with the combine, the spout dead centre on the trailer, when I hear Roger screaming at me from his cab, and peeping the horn. In all the hurry I'd forgotten to fasten down the tailgate on the bleedin' trailer, and the corn was pouring out the back as quick as it was coming in. There was at least a ton of the stuff in a twenty-yard trail behind me.

I considered bribing Roger to keep quiet about it, but figured that would just be more money down the drain.

He finished the field at nine o'clock, and after everybody had gone home I went back in the dark and shovelled up the mess. Gladys was upset 'cos m' supper was spoiled by the time I got in.

Saturday August 8th

A canny day. Baling straw with Charlie's big round baler, — the machine gobbles it up. Gordon took over after dinner and I took a barley sample to Northern Grain to see what it was worth. The reckon the moisture content is 18 per cent so it doesn't matter what any other meter says, — and the nitrogen content is far too high for malting. They also tell me that, sadly, the feed price is depressed (it always *is* by the time I come to the market) but they generously offer to take it off my hands, with appropriate deductions for moisture, within the next day or two.

Willie played cricket, and scored seventeen (top scorer). He said half the team were missing, harvesting. They had four schoolkids playing and an OAP. They lost again.

Sunday August 9th

Finished baling straw without a disaster (this is another record!) Took the baler back to Charlie's, very relieved at no breakages.

Began leading bales after dinner, but Willie fancied the job, so as we're just piling them up behind the hayshed, I let him get on with it. If he's not going to be an academic, he might as well be useful.

Wayne came for supper, and he and Doreen snogged in the kitchen all night. Willie sneaked in for a biscuit every twenty minutes and came back to the telly with a progress report. I think Doreen may kill her brother before he's much older.

Monday August 10th

Wagon in for barley at seven. (I'm just up in time, thank god.) Had to hurriedly replace the fork with the bucket, while the driver paced up and down the shed. He says he'll be back again after dinner.

He's not of course. He drives in just as I'm sitting down for m' dinner, and it takes me until half-past one to get him loaded. He has the ticket for the first one, just over twenty-two tons, so it's weighing quite well. Is there a third load left? Maybe. If there is, that would make it nearly three tons an acre! Stop it, − don't count it up till you've seen the official weights, and remember there's off-takes for moisture.

Finished leading the bales, by which time it was raining.

Tuesday August 11th

Wet. No wagon, − phoned Northern Grain who say they won't get back today.

Haven't looked at the stock properly for a while, so take more time over the job this morning. Ewes all OK. Cattle seem to be doing well, some lambs scoured, and a lot of lame sheep hobbling about. What is it about sheep's feet? They're always rotten!

Wandered through the next field of winter barley, − and panic. It's ripe! Even a few heads off!

A wagon comes for the other load while I'm miles away from the farm (I thought they weren't coming back today!)

Phone Roger after supper, trying to inject some urgency into my voice, — it isn't difficult.

Wednesday August 12th

It's still raining, which means no combining here, or anywhere else. It also means that Roger will be delayed for another day at least, after the rain stops. (It will stop, won't it?)

Go up to the mart for a look at the trade, and m' dinner. The trade is a 'flyer' (it would be of course, when I've got nothing there). Otherwise everybody is pretty miserable about the weather. Even Cecil is less than his usual boring chirpy self, because he's got twenty acres of rape lying in soggy swathes.

Thursday August 13th

It's not actually raining, but you get the impression that it would if it could. Meanwhile the reps have worked out that all the farmers will be trapped in their farmhouses today, with nothing to do, — so they're out in force in their company Cavaliers. Some of them aren't much older than Willie, but they've read the literature, been on the weekend course, and now they're experts. My rejections begin with, 'no thank you'. Progress through 'never use the stuff', to a slammed door, and 'why don't you bugger off bonny lad ... ?' After that I pretend not to be at home, hiding along the passage, in case one of them looks through a window and spies me watching the racing on telly. I am eventually 'caught' by the man from Northern Grain. Stupidly imagining he might have a barley cheque about his person, I invited him in and gave him a cup of tea. He left with an order for winter seed, but only on the understanding that the bill wouldn't be paid until December.

Friday August 14th

An unsettled day with scattered showers. That's an improvement. The townie reader might get the impression we're completely obsessed with the weather, — and it's true enough. If you work in an office, a factory or a classroom, — it's hardly critical if it's peeing down outside is it? A cloudy day won't affect the pay packet or the index-linked pension!

Dosed the scoured lambs, and put them through the footbath.

Gladys is away shopping for Willie's wardrobe. She says he's just grown out of everything, can't get into his shoes, trousers half-way up his legs, arms sticking out of his jacket, shirts bursting at the seams. I suppose it's inevitable. He certainly eats more than the rest of the household put together. Feeding him is like feeding the thresher, — whoomf, and it's gone!

Talking of food, Florrie cooks m' dinner today, and that's always a special event. She's so enthusiastic, so delighted to be given the job once in a while, you feel you have to eat it all, give her a clean plate, generous compliments, — but it's not easy. In fact it's always a disaster, she cremates everything.

She's a 'cordon-noir' cook, — it's inedible! When I know she's
preparing the dinner I generally invite Sweep to help me out,
but he's not impressed either.

A phone call at night confirms that Roger didn't combine
anything today.

Saturday August 15th

A howling gale. I daren't look at the barley, all the heads will
be off by now for sure. And the rape, oh my god, — the pods
will be empty, — we're ruined, penniless, debts galore, wife
and children destitute. Montague will send in the receiver.
I may hang m'self in the stable!

Alright, I'll have m' dinner first, then twenty minutes kip,
and see what I feel like after that. . . . Worse! it's blowing
a hurricane. I set off to see how Roger's getting on.

Charlie is trying to bale straw, but he can't catch it.

Cecil Potts is combining his rape, but half of it is missing
the trailer.

Arthur is ploughing stubble (maybe I should be doing that).

Roger is combining in a cloud of straw and dust, — 'says
he should get to me tomorrow. But will we have anything left
by then?

Sunday August 16th

Roger arrives at noon, red-eyed from the dust and wind, and
having combined through to midnight. We start our field at
one o'clock. He wants it finished tonight because he's
promised someone else he'll be there in the morning. (Con-
tractors are superb promisers.) He goes flat out (how much
is coming over the back?) I'm not going to look, just drive
the tractor, get in the right place under the spout, take it home,
tip it, and hurry back for the next load. Don't even think!

I recall Arthur Thompson once had a field of barley, a lot

of which had snapped off. The fool went and counted the heads in a measured square yard to assess what weight of corn had fallen onto the ground. He would've felt much better if he hadn't bothered.

Roger finished at ten o'clock, had a cup of tea in the kitchen, and went home in the pickup. I had a bath, and went to bed, — knackered with nervous exhaustion.

Monday August 17th

The wind has died down, and it's raining again. For some reason I feel quite pleased about that, and just a little ashamed that I feel pleased. I suppose it's all because we managed to get a field finished, while some other poor sod is probably stuck in the middle, or didn't even get started . . . and they won't cut anything today.

Gladys is in a bad fettle because she cannot hang the washing out (apparently this is critical). Florrie blames Chernobyl.

Tuesday August 18th

A bad day. Wagon came in for barley, and I drove the bucket into the side of his cab. The driver was quite upset. Then a puncture in the front wheel of the tractor with only half the barley loaded. To the garage in haste to get it repaired, and by the time I got back the wagon driver had rigged up an auger to finish the job. This was good thinking, and showed considerable initiative, but unfortunately he got the shovel caught in the worm at the bottom, and before he could stop the auger, the belt had burned off at the motor, and the lights were all fused.

Gladys came round to tell us her hoover had stopped (she should've known better).

Wednesday August 19th

Lamb dead in the hay field ... Pulpy kidney? Pneumonia? Bubonic plague? Dandruff? ... who knows. What's that old saying about never counting your chickens till they're hatched? Well you should never count your sheep till the auctioneer's hammer comes down. Perhaps you shouldn't count sheep at all.

The wagon comes back for barley with a very nervous driver, but this time he's loaded and away in forty-five minutes, relief on both sides.

Took four bullocks to the 'fat', − all graded, but one behaved very badly. This one had always been a bit excitable, sometimes positively deranged, and today was his big day. It was also his last!

For starters he was reluctant to come out of the field, and Willie and I ran about in rings trying to get the beast out, until I was on the verge of collapse. From then on the performance went thus: with goggly eyes and tongue hanging out he ignored the open gate, jumped the fence onto the road, and galloped off in the wrong direction. Chased him in the pickup and turned him into the Pillicks' garden. Out of garden over designer wall, removing several startled gnomes, and

back home, where Gladys and Jake the postie bravely turned him into the yard. He was then 'persuaded' into the byre with his mates (who had walked quietly back to the farm without any fuss). By this time the wagon was waiting.

He got half-way up the tailboard, then jumped sideways and smashed the gates on the wagon. We had to let the others out again, tie up new gates, surround the loading area with men, women and obsolete machinery, — and then, screaming like drunk Apaches, we managed to chase the whole lot into the wagon in one made rush, and whip up the tailboard. Even inside the wagon the fool continued to bang and rattle all the way to the mart. By then, exhausted and undoubtedly several kilos lighter, he went to his appointed pen without much trouble. But by the time the grader got to him the beast had his third wind, and Ghadaffi thought it prudent to grade him from the alleyway.

Eventually to the ring at full throttle, bystanders taking refuge in empty pens. There followed a brave attempt to demolish the weighbridge, the rapid dispersal of over-weight butchers, and a strategic withdrawal to the back of the gallery by the whole congregation, as the bullock made one circuit and out into the daylight again. McMurdie shouted 'stand on', and knocked him down immediately to the last buyer, who was nowhere to be seen.

Baled straw peacefully when I got home.

Thursday August 20th

The Pillicks are disappointed about the state of their garden. It's definitely an insurance job and, having paid millions in premiums over several decades, I have no qualms about claiming a wee bit back.

Ordered more seed and fertiliser for autumn sowing. Ye gods, the income barely pauses long enough to make my acquaintance before it leaves hurriedly on another mission.

Finished baling, and Willie led a few. The forecast is for rain.

Friday August 21st

McCaskill was right, it's a soggy summer's day.

Roger's man came to plough the barley stubble with a tractor the size of a three-bedroom house and a five-furrow reversible that turned over the whole field in less than a day. It was as if a giant had picked up the corners of a tablecloth and flicked it over. The operation would've taken me most of the week, and *then* the settings out would be wonky, the finishings triangular, the breakages frequent, and the temper short. It doesn't take long to convince m'self that Roger's investment makes more sense than mine, — especially with Montague breathing down my overdraft.

I led more bales in the other field, and in between journeys forced myself to look at the rape. Maybe this wasn't such a good idea, — I get the impression that if I break wind, the whole fifteen acres will fall out.

Saturday August 22nd

Gladys says if I'm not combining, how about a nice run out in the car ... ? So after dinner I take her to inspect the rape again, have a quick look at the spring barley and the winter wheat, see if there's any cattle fit enough for Wednesday, and take a rape sample down to Northern Grain. It's 20 per cent.

She never said a word the whole trip. You just can't please some women, can y' ... ?

The rape problem is straightforward. The current moisture content is too high and too expensive, — but if we wait too long (and get a shake wind) there'll be nowt left. Futhermore the price is falling daily. Do I need these problems?

Gladys reminds me that it's Doreen's birthday next week, and tells me I've already bought a bracelet costing thirty quid (really?) and can she have the money now please. What's more the girl is having a party next Saturday night, and we'll have to go elsewhere. My lifestyle is disintegrating! Don't forget Gladys's birthday for God's sake!

Sunday August 23rd

According to Arthur's meter the rape moisture is falling. The wind, on the other hand, is rising. I'm eating cigarettes, I'm speaking to nobody but m'self. I promise God I'll talk to Gladys's mother, give a fiver to the RSPCA, stop swearing in a very loud voice at the dog, change m' wellie socks at the end of the week, — *anything,* if he promises not to blow a gale.

Phone Roger after supper to order the combine. 'Immediately will do nicely,' I tell him, trying not to give the impression that I'm a jibbering wreck.

Tuesday at the earliest, he says.

Monday August 24th

A breezy day. In any other circumstances it would be a good 'n'. Everybody else seems excessively full of the joys, glad to be alive. The postie is whistling merrily, the milkman actually breaks into song, Willie has a silly grin on his face, in spite of having been ordered to cut the lawn. Doreen goes off to work with renewed enthusiasm dressed in various birthday presents. Sweep lies in the sun watching Willie, and wagging his tail. Gladys is doing a foxtrot with the vacuum cleaner.

And all the time the oilseed rape is rattling to the floor, and ruin stares me in the face.

After the nine o'clock news McCaskill's isobars are still too close for comfort.

Tuesday August 25th

Roger phones at breakfast time to say he's still got twenty-five acres to cut at Fletcher's, and might not reach me today. He's had a breakdown! Briefly consider both suicide and murder, — but set off around the stock instead, muttering. Sweep, who can recognise stress in humans, trots along behind, just out of stick-throwing range.

Later I lead the rest of the straw in bright sunshine.

A restless night, — haven't slept properly for a week. Gladys goes out like a zombie, and gives her usual impression of nearby thunder.

Wednesday August 26th

The forecast is for showers later in the day. The fine spell may be breaking up. Prayers are offered for just another twenty-four hours. God must think I'm a right creep, — but the situation is desperate, and being nice to him surely can't do any harm.

Roger's machine looms into view at ten o'clock. It seems to take him ages to grease the thing, fill up the diesel tank, change the riddles, the cutter bar, and have his bloody coffee, — but we get going at eleven.

We have a wagon trailer standing in the field. With this creepy-crawly rape the container has to be watertight, and my old tractor trailer is hardly that.

It's all too obvious by the time it takes Roger to get a tankful, that there's a canny bit lying on the ground. Nowt we can do about it. Just don't look, stay out of the field, — tomorrow could be worse!

Thursday August 27th

Roger finished and left at about seven o'clock yesterday. The results from Northern Grain today are not exciting. The good news is that the moisture content was quite low and the admix rubbish was quite low, — but so was the yield. Just over a ton an acre.

I won't grow the stuff again. I've aged a decade in a week. In fact if I could afford it, I'd just be a dog 'n' stick man. It's got to be a less traumatic way to farm ... and you can't beat that satisfying feeling when you walk round the stock on a bonny night, and see them all grazin' or lying, contendedly chewin' the cud. Maybe humans would be calmer if *we* chewed the cud. . . .

Friday August 28th

It's a bright morning as I walk out to inspect the stock and, surprise surprise, I come across a recently deceased lamb (what was I saying yesterday . . . ?) And no, it's not the rubbishy little crit I half-expected to snuff it, it's not the one who's been scoured skinny since birth, nor that pathetic triplet that's never had a decent suck since he was born. It is of course a superb creature who was destined for the mart next week. The sod!

To the pub, where I add a couple of mythical hundredweights to the rape yield, but still fail to impress anybody.

Charlie says he had a dead lamb this morning. I express deep sympathy, but I don't mention mine.

Saturday August 29th

Preparations for Doreen's party are in full swing from dawn. There's nowhere to sit down, wellies have to be left several miles from the house, the hoover is unstoppable, the cooker is churning out sausage rolls and vol-au-vents, the telephone is no sooner put back on its cradle than it cries to be picked up again, a van delivers enough booze to satisfy the Hunt Ball, even Florrie has been persuaded in for half a day to help (hopefully she won't be allowed to cook anything). I get a cup of instant soup for m' dinner.

Willie wisely leaves early for his cricket match.

The bright young things begin arriving as Gladys and I leave. I get the distinct impression that they're all grinning in anticipation of an orgy. I have visions of naked cocaine-sniffing countryfolk being sick all over our house, — but Gladys drags me away for supper at the Queens. (She calls it dinner.)

It's all very sophisticated, drinkies in the bar, nuts and crisps, and a poncy waiter brings a massive menu in a foreign language. I stick to broth and a sirloin steak, but Gladys is easily carried away on these occassions, and stuffs herself with garlic prawns, and some other fancy fish dish covered in alcoholic gravy.

She has her back to the wall, as usual, so she can see what's goin' on, — and gets her eye on Humphrey Smith, the destitute wheeler-dealer, over on a corner table, blathering intimately with a lady. He's obviously not as hard up as we've been led to believe, or he couldn't afford to eat here, — and that can't be his wife either, or they wouldn't be talking so much.

We call in for a nightcap at Charlie's (Gladys said it was still too early to go home) and eventually get back at two in the morning.

There are still several cars in the yard, but the house is in total darkness. Gladys opens the back door as noisily as she can, rattles the sneck, and disintegrates into an unconvincing bout of whooping cough. But sure enough this dramatic performance has the desired effect (no fool, that woman). The lights come on, dishevelled couples creep out from all over the building, smiling sheepishly (though to be honest I've never seen a sheep smile).

'Did you have a nice time?' asks Doreen after she's pushed Wayne into Daddy's Range Rover.

'Lovely,' says Gladys, . . . 'there seems to be a lot of food left. . . . ' But as I told her, this lot didn't come to nibble vol-au-vents!

Sunday August 30th

There's nowhere to sit. The hoover is out of control again, — it's tidying-up day. Inquests are held, characters are being destroyed, mother and daughter are in full flow, — and the dinner's delayed until 2.00 pm, a townie lunchtime!

Outside there's peace and a nice breeze, — there's also some lambs through into the spring barley. It's undersown and they're trampling the barley to gorge themselves on the clover. Sweep and I get them out and mend the fence. (Well actually Sweep wasn't a lot of help with the second job. . . .)

The barley's ripe, necked, beginning to bend over into the seeds. Phone Roger after supper. The house is more or less back to normal, but you can still smell there's been a party.

Monday August 31st   Bank Holiday

A fine quiet day, — decide to burn the rape trash.

I do everything right, according to the code. Move the outside swathes, plough around the headland, stick a wet finger into the air, and set it alight. At this very moment a force-nine gale sneaks up behind me, and within minutes the flames are away up the field totally out of control. I have to move quickly to get the tractor and the little three-furrow plough to safety. Then all I can do is watch. It's all gone in an hour, — whoosh, charred stubble, black ash flying east. A highly successful burn.

Twenty minutes later I'm having a cup of tea, and Mrs Pillick comes to the door looking like a dis-chuffed Whoopie Goldburg. She is very upset that her washing, her windows, her gnomes and her pastel yellow three-piece suite are all much darker now. She says I'm a menace to the community, and the police have been informed.

Well, how was I to know it was *another* bleedin' Bank Holiday. I'll have to plead insanity.

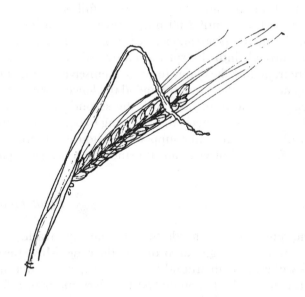

September

Tuesday September 1st

Combining the spring barley, — a moderate crop. Roger is cutting some heads off as he claws his way through the rampant clover. Moisture content is not much above 17 per cent, — according to my teeth.

Willie leads off without a mishap, — until the very last load. It was then that Tracey came and joined him in the cab. Willie's concentration lapsed, and he removed the stackyard gate post.

Wednesday September 2nd

A wet drizzling day, damn it. I was hoping to get the straw baled on the seeds. Took some lambs to the mart instead. Everybody else is obviously thinking along the same lines, and the pens are awash with sheep. The trade is depressed, and we all hope it's depressed all over the country, so that the subsidy makes up the drop.

In the canteen we compare harvesting notes. Charlie had a breakdown yesterday, and still has twenty acres of barley to cut. Cecil is into some winter wheat. Fletcher says his yields are pathetic, and he's probably ruined (but he always talks like that, then goes to Spain to play golf). Arthur says he's got a field of wheat dying rapidly from some disease (so what's new?) Percy has seven lambs rejected, and reckons Ghadaffi should be sent to the Falklands.

Willie and Tracey are at the mart, and Tracey is blamed

for the poor trade. McMurdie says nobody's concentrating. He could be right. It's always the same when a bonny lass appears at a mart, — the peasants suddenly behave like randy trappist monks who haven't seen a female for years!

Thursday September 3rd

A better day, — we've got to get that straw baled, so I persuade Willie to wuffle it while I cultivate for oilseed rape. (Yes, I know I wasn't going to grow the stuff again, but that was *last* week.)

Roger says he can sow the rape on Saturday. It's all go, — we haven't got this harvest yet, and here we are sowing the next one. It didn't used to be like this, — autumn was once a time for relaxation, contemplation, — mellow fruitfulness and all that. Now it's one mad frenetic dash to beat the winter.

Winter? — sshh, — don't talk about it.

Friday September 4th

Grab your chance, it's hardly perfect, but we'll bale that straw.

I do the baling, Willie goes on the cultivator. The land's breaking down nicely, should be a canny seed bed by the darkenin'.

Down to the Swan after supper, fingers crossed for another dry day tomorrow. Gordon (the local weather expert) has seen two snails mating on top of a fence, and this is a sure sign of fine weather . . . he says.

Saturday September 5th

Gordon's right, — but I still spend a nervous morning waiting for Roger's drill, — trailer poised loaded with seed and fertiliser. In fact he doesn't appear till after dinner, by which time I'm doing a lot of aimless pacing and chain smoking, convinced every whiff of cloud heralds a monsoon. However it's all sown and rolled in by nightfall.

Willie plays his last game of cricket, and ends the season in a blaze of indifference, — three runs and a catch.

Sunday September 6th

A brilliant day. Everybody is combining except us again, — we've got nothing ready. The wheat's *nearly* ripe, maybe a week away yet.

I lead bales off the seeds, while Doreen and Tracey sunbathe in the garden, and Willie makes a superb job of cutting the lawn, very carefully.

A warm balmy night, — midgies dancin' at the back door.

Monday September 7th

Son and heir is back to school, with new shoes, pants, sweater, socks, and a threat from Gladys that if he doesn't work harder he'll just end up like his father! I get the impression this prospect is considered only slightly more attractive than becoming a destitute child molester.

And it's pouring, the midgies are all drowned, the spuggies are sheltering in the grain shed. Nothing we can do in this weather, — went to Charlie's for coffee and a chat. He's bought a new collie dog (Moss is getting past it, poor thing) but it's giving him some trouble. He can't get it to run at all. He's tried everything, whistling, swearing, screaming. The dog follows him everywhere he goes, won't leave his side, just sits there smiling and wagging his tail, while Charlie jumps up and down.

Hilda says it'll be the death of him, — but I suspect it's the dog that's at risk.

Tuesday September 8th

Go to a sheep sale to buy a tup and some gimmers. Everything is very dear. It's another wet day and the entire farming population has nothing better to do than spend money. Everybody's there, waving their cheque books, — we're all mad. What was that quotation I heard the other day, something like, 'farming is the only business that buys retail, sells wholesale, and pays the haulage both ways!'

I tried for two of Jimmy Fletcher's Suffolk shearlings, but was runner-up to Cecil each time. Maybe I should've gone on, but I think Cecil would've bid forever. Eventually bought a canny lookin' beast for about twenty quid more than it should've been. — Never mind he's got plenty bone, and a good head, — he'll be alright.

Didn't get any gimmers, — they were a ridiculous trade, but got fifteen one-crop mules from a bloke going out of sheep altogether (what does he know that we don't?)

Cecil says his tups are for nowt. When mine came home

I locked him up in a very confined space in the byre with the 'old boys', so they could get acquainted and discuss tactics.

Wednesday September 9th

Would you believe it, it's still raining, — we're at a standstill, can't sow barley, can't combine wheat, can't do anything useful. Everybody's in the same boat I suppose, but it's still very frustrating. And some of the wheat's beginning to shed.

Charlie's dog has moved. Last night, threatened with a twelve bore, he took off like a bullet, and chased the sheep round and round the field. Several aged yowes may not recover. Charlie may not recover either. Hilda says he's just sitting in the kitchen moaning and swearing at the same time. The dog's sitting at the back door smiling and wagging his tail.

Thursday September 10th

It's fair, and McCaskill is promising some further improvement over the next few days. Gordon, presumably having observed a hedgehog dancing the paso doble (or some such 'natural' phenomenon) is also convinced the weather's on the mend.

Friday September 11th

Walter Williamson drops in with the good news that our wheat is riddled with eye-spot and take-all. I give him a cup of coffee and tell him to go away.

The ground is very wet now, and a rubbed-out sample of wheat looks like damp brown rice. There's no drying, just a muggy quiet day.

Went down to the pub where miserable peasants grumbled their way through several pints, while the ladies dart team screeched their wobbly way to victory over the Plough.

Heard that one of Cecil's expensive new tups has snuffed it. (Unlucky.) Apparently the crazy macho beast, having first tried to beat the living daylights out of his mate, then challenged a Massey Ferguson 3000, and came second.

Saturday September 12th

Boadicea comes for the weekend and, with Willie away playing rugby, the house is full of blathering women. The auld granny seems to generate more hot air than a grain dryer (perhaps we could harness her to a massive fan, with ducts leading off under the floor?) Gladys and Doreen are little better when she's around.

I wish we were combining, or working land, or doing something outside. In the end, after spending an hour talking to Sweep in the stable, I take the link box and spend the afternoon mending a few dodgy fences. − Peace, tranquillity, only m'self to talk to, − until Charlie comes over the hill with his new wonder dog.

Flash (that's his name would y' believe) has got the message at last. He now gathers everything at breakneck speed, − sheep, cattle, hens, ducks, everything that cannot actually fly is brought to Charlie, generally in a state of collapse.

Charlie and I are having a quiet fag, and before we realise it this over-enthusiastic mut has gathered all the sheep in my field, and they're standing there gasping. Sweep will have nothing to do with the fool. I think he's embarrassed.

Sunday September 13th

A better day, brighter, harder, just the occasional light shower. There are a few desperate combines going, and Cecil is sowing winter barley on some light land.

The womenfolk go out in the car for tea somewhere, so I have a lazy afternoon reading the papers. The phone goes twice, but I don't answer the damn thing.

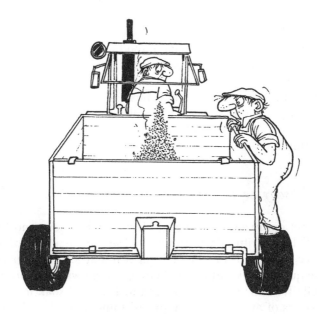

Monday September 14th

Roger phones at breakfast (says he tried several times yesterday but got no reply) to tell me he'll be in to cut wheat tonight.

Panic, I'm not sure I want him yet. It's still as soft as clarts, chewey. It's not ripe enough really, but if I put him off, he might not come back again for ages.

Fortunately it's a good windy day, it *will* be drying. Relax, he'll probably be a day late anyway.

He drives in after tea, says he's come from Arthur's where the stuff is 30 per cent. I have to tell him mine might not be a lot better, but we'll have a go.

Twenty yards up the dyke a belt snaps, and he can't get a replacement until tomorrow. This (I conclude) is not a disaster. It will delay him for a few more drying hours, but it keeps him here. The little bit we've got in the trailer is standing up like Cleopatra's needle.

Tuesday September 15th

We're in luck. There's been a good airy night, and it's still fine and breezy, — the wheat could be two or three per cent lower than yesterday.

Roger has a new belt, but he has to dismantle half the combine to get it on. It takes him till dinner time.

Once around the headland, and we've got a tankful. And it's raining!

Wednesday September 16th

We get going again on a cool blustery afternoon, and finish the field well after dark. In the shed we've got a heap of cold, sticky shrivelled-up wheat. I guess it's at least 25 per cent. The quicker it's shifted the better, it'll be warm in no time.

Gladys runs me a hot bath, and I lie and soak for three-quarters of an hour with a gin 'n' tonic and a fag. If God's any sort of bloke at all, heaven will be like this.

Thursday September 17th

I can't get any wagons until tomorrow, and the sample I take to Northern Grain reads 23 per cent, — better than I thought.

Haven't looked at the stock properly for a day or two. I've just had a quick run round in the pickup to ensure that most of them are in the right place, and the right way up. I find the yowes have wandered through onto the wheat stubble (we left a gate open late last night). Have they been there long? Have they stuffed themselves with wheat? Will they all die a ghastly gastric death?

Sweep gets them out with Flash-like urgency. Is it my imagination, or are some of them twice their normal size, and staggering a bit? I check them again at night. A few of them are scittered, but so far, so good.

Friday September 18th

The wagon's in at 7.00 am. Willie loads one effortlessly before the school bus comes. I load the second, and take half an hour longer, spilling wheat all over the yard. The pullets think that's great.

Breakfast at nine with Florrie mopping the floor around my feet, and telling me about the widow Simpson's latest affair. 'As you know,' she says, in her posh voice, 'I was never one for unconfined gossip . . . but . . . ' and off she goes with some sordid little tale of village life.

A third wagon sails in mid morning. By the time he leaves there's not much wheat left, hardly a load, and it's heating up already. I shovel it sideways to let the air in.

The yowes are still alive when I eventually see them, but one is looking a bit wobbly. I remember something about dehydration in these cases, and pour some glucose and water down her throat. She pretends to choke to death.

Saturday September 19th

Cultivating land for winter barley. Of course the seed hasn't arrived yet (this happens every year) but they promise it will definitely be delivered today. Roger says the drill is available whenever I'm ready.

By nightfall, still no seed, and I'm cursing everybody.

Willie comes home late from rugby with a black eye. They won 12 − 3.

Sunday September 20th

Seed delivered at dinner time. The driver is full of excuses. 'Not my fault' (never is), but I suspect he holed in at some hostelry last night, and had a lie in this morning.

Willie and I took turns on the cultivator, and got all the

barley land ready for sowing tomorrow. I hope it doesn't rain tonight, or we'll have forty acres of porridge.

Doreen is depressed. The moody misery is moping about as if she carried the cares of the Common Market on her shoulders! What's she got to worry about? Gets a wage every month, holidays with pay, — doesn't matter to her if it's wet or Wednesday. . . . I feel like giving her a good spankin', but I suppose I'd get arrested for that now.

Then Gladys explains to me that Wayne is 'off the team'. So bleedin' what, he was a pain anyway, any son of Cecil *would* be. Doreen says he's a pompous, boring git, and she's sick of 'im. So what's she so depressed about, for God's sake? Women are a mystery.

Monday September 21st

All day it's corn-sowing action. Pre-emergent spray, courtesy of Walter's company; drilling by Roger; provider of seed, fertiliser and slug pellets, m'self. What an investment. It's a lot of pounds per acre, and it'll lie there for a whole year, while Montague and I worry about it.

It's windy. The empty plastic bags are dancing off towards the village, but most of 'em stick in the hedge. By suppertime all the flotsam is gathered up, and the farm is quiet again.

Tuesday September 22nd

The greedy yow who gorged herself on the wheat stubble is a gonna (three-crop, probably worth eighty quid).

I still need a few replacement breeding sheep, and there's a sale today, − but the weather's good, so I decide to burn wheat straw. We don't need it for bedding, so I only bale the headland, lead it off, plough, and set fire to the rest. That all sounds very simple, but for a mechanical moron like me, who has difficulty connecting implements to tractors, never mind operating them properly, − the operation takes all day.

In the evening I move that little pile of warm wheat again, and urge Northern Grain to shift it soon.

Wednesday September 23rd

Both cattle and sheep to the mart. Ghadaffi is in this most arrogant mood, rejecting lambs from almost every pen. We have a couple who get the purple mark of disapproval (they also get a reprieve, and go home).

It's a long, weary thirty-cigarette day, from lambs weighed to cattle sold. I'm early on the ballot for sheep, and nearly last with the bullocks. By mid-afternoon the cups of coffee are sloshing about inside, the beef sandwich lies like a breeze block, and m' mouth tastes like an ash tray. Charlie and Arthur go to the pub, but that's no good to me. Half a shandy in the middle of the day, and I'll be nodding off.

Humphrey Smith is back in business, cockier than ever. He's not buying for himself any more though, but rather for some big wholesale outfit which specialises in redundant cows, sows and yowes for the connoisseurs of Bradford.

179

Thursday September 24th

A wagon comes in for the rest of the wheat. The sweepin's-up are kept for our modest poultry flock.

Twenty acres of wheat still to combine, and we just need a few good days for that to harden, a touch of frost perhaps.

Borrow Charlie's Japanese trike and spinner, and scatter slug pellets on some of the newly sown corn. Never handled one of these things before, and I feel like Barry Sheene. Should I be wearing goggles and all the flash leather gear, and do 'wheelies' at the end of the field? Jesus it's cold. Couldn't they put a little cab on this machine?

One of Roger's lads is ploughing the burnt wheat stubble. There was a shower last night, so the ash is still.

Friday September 25th

Hurricane Harriet arrived in the early hours of this morning! She woke me up about four o'clock belting on the windows and rattling the doors. At five I got up to see what she was doing. Opened the back door, and she lifted the lino right along the passage. There are slates off the roof, water lying everywhere. The early news talks of wagons cowped on the motorways, fallen trees, extensive flooding, train services disrupted, electricity lines down, and general chaos in the city. It makes no mention of my wheat! The wheat field is a disaster area. It looks like a flat ocean, disturbed only by regular lines of waves that don't move. And yesterday it looked great.

At the pub we compare disasters. Arthur has twenty acres of oats that were six-foot high yesterday; now you could see a mouse playing at the other end of the field. Charlie has a lot of wheat like mine, and a tree fell on a yow. Jimmy Fletcher had a combine in the middle of a field of Avalon, and it's stuck there.

Davy Scott lost the roof off his grain shed; someone else had a hen house demolished and spent all day catching bewildered pullets. Percy has a lot of flat corn, and lost a field of straw completely. He feels strongly that McCaskill (who forecast a slight frost) should be taken for a ride inside a rotaspreader, before being hanged with baler twine from the delivery spout of a combine.

Cecil, as you might expect, finished his harvest two days ago, and nobody's talking to him!

Saturday September 26th

Harriet has gone off to ravage Scandinavia, leaving behind a desolate grey countryside, full of disillusioned peasants. What a way to nearly make a living! None of us dare work out what the last twenty-four hours might have cost.

At night Gladys and I are invited to another party at the Pillicks'. Personally I don't want to go, I'd prefer to stay at home, quiet and miserable, — but Gladys says it'll do me good, take m' mind off things, and we'll meet different people.

They're different alright. Nobody's interested in the price of mule yowes or barley yields, or how Harriet stole my wheat away. All their chat is about their own enormous salaries, colossal pensions, rising property values, holidays on yachts, and how brilliantly their spoiled little brats are doing at prep school. I'm out of m' depth again, and sneak away at eleven o'clock, explaining to Pillick that we've got a heifer calving. Gladys gives me one of her auld-fashioned looks. We're fresh out of heifers of course, — but Pillick wouldn't know the difference between a heifer and a haemorrhoid!

Sunday September 27th

The wheat's a mess. In the old binder days we'd never be able to cut it, — but with Roger's clever machine we might still chew our way through most of it, — when it dries out.

Meanwhile Doreen, devastated by the loss of Wayne the

wimp, has found consolation in Roderick the randy. This young man has apparently been eyeing her up every time she put her wages into the Building Society, where he works. Needless to say he is not of sound peasant stock (nobody's perfect) but to be fair he seems more or less normal, even if he *has* got very small hands and clean finger nails.

Gladys approves because he's tidy. More important perhaps, so does Willie, because the bloke plays rugby and can run like the wind, though obviously not fast enough to avoid Doreen's devastating tackle.

Monday September 28th

It's a day for cleaning up after the storm. Phone Newcassel-Browne to discuss repairs, but he's had a lot of similar calls, and is in a very grumpy fettle. What the hell does he expect? Does he think that after giving us a rent rise, we miserable tenants are now being ungrateful by asking for a roof over our heads?

There's an old ash tree down in the back field, − it's up by the roots and it's taken a stretch of fencing with it. I take the tractor, the link box, posts, nails and chainsaw to repair it, − before the bullocks find the hole and jump through. Curiosity demands they come and watch, and sniff and lick. Cattle are like that.

Tuesday September 29th

We still can't contemplate any arable farming. The wheat is still sodden, and the ground is too. So I get the yowes into the pens to dock tails and clean them up for their honeymoon. Nobody but an umpteenth-generation peasant would consider spending any time cleaning the rear end of a sheep, who has quite obviously been more than 'regular'.

I pare their toe nails as well, then beat them mercilessly through the footbath. By the end of the day I'm covered in blood and muck.

Wednesday September 30th

Found five dodgy ladies among the ewes yesterday that really shouldn't be kept. They'll never survive a bleak week in January, and go on to feed lambs. So they go to the mart, along with a dozen lambs and four bullocks.

The old yowes are a 'flyer'. (Humphrey buys them, and when I give him some luck money, he just smiles.) The lambs were all graded, though I thought one was pretty lean really — Ghadaffi knows nowt. The cattle were all sold, but certainly didn't leave a fortune. That's the problem with cattle, the investment is enormous, and the interest often less than the Halifax.

Arthur Thompson topped the mart with a superb Charolais beast. When I congratulated him later in the canteen, he told me (very quietly) that it actually lost a fiver.

October

Thursday October 1st

The countryside is drying out at last, a fine breezy day. Roger has dragged his combine out of the wet hole in Fletcher's field and is combining slowly.

I begin ploughing slowly with m' little three-furrow toy, and a million seagulls appear as if by magic. It takes me all morning to set out, mark the headland and get the plough running properly, — but by tea-time I've blacked over a few acres and made a right mess of two finishin's.

The seagulls leave at five o'clock, apparently unimpressed.

Friday October 2nd

Whoopee! Fantastic, tremendous!

Gladys is bewildered by my good humour (indeed by anything more demonstrative than a cough so early in the morning). Sweep grins and wags his tail, he knows when I'm in a good mood. Florrie is suspicious, and stops vacuuming to suggest I see a physiotherapist (I think she means a psychiatrist), — but no matter. Jake the postie has delivered glad tidings of comfort and joy, — namely two big cheques, one for grain, the other from the mart.

To the bank at 9.29 am.

Saturday October 3rd

I'd almost forgotten about the 7th cavalry, but they've started hunting again, — they're out 'cubbing' in the early morn. Knickers, Newcassel-Browne and a small posse of the faithful come charging over the stubble while I'm in the middle of a tricky finishing (ten yards wide at one end, and maybe a foot at the other).

The Colonel makes his enlightened landlord/Commander-in-Chief speech. He calls me by my surname (no mister), probably a hangover from Eton and Sandhurst. ('I say Tomkins, Smithers, Fontleroy,' or whoever, — 'Barclay's pinched my tuck box'.) Anyway he bellows on for a few minutes from his perch on his demented gelding, pretending he knows something about wet wheat and dead yowes, the Ayatollah hovering t' heel. He tells me how the hurricane blew over his Victorian gazebo, then finally says, 'jolly good, — carry on . . . !' touches his hat with his riding crop, and fearlessly gallops off to face the foe. I am dismissed, like a petrified private on the eve of Waterloo.

Sunday October 4th

I think the wheat will cut soon. The ground's still wet, but the straw is drying out. Roger is combining Percy's flattened field, — it's a very slow job.

Willie is feeling fragile. Yesterday's game was against a team of neolithic police cadets. Whenever our lot won the ball in the scrum, which wasn't very often, — they were always reversing at speed, consequently the scrum half (Willie) was repeatedly submerged. He's unusually quiet. Tracey comes to comfort him on the sofa, while I finish the ploughing.

Gladys and Doreen go to the Harvest Thanksgiving. That dosy old vicar always arranges it before we're finished. How the hell can you sing 'All is safely gathered in . . . ' when there's still acres of black wheat lying in the fields? Some day,

some desperate peasant with a knackered combine and twenty acres still to cut will kill the bloke!

Roderick came for supper. I reckon he's a bit of a lad, — Doreen couldn't take her eyes off 'im (and I think he was up to no good under the table). He was extremely polite to Gladys (he's no fool either) but I figure he went too far when he helped with the washing up!

Monday October 5th

A howling gale, — this is just what we need, there's certainly no fear of the wheat blowing out.

Cowed the last of the yowes, and went through the lambs looking for any fit enough for Wednesday, and marked a dozen. We'll also get a few more bullocks away. There's loads of grass still. We should be buying some cattle, and lambs for the turnips as well. It's just a carefree money-go-round.

Never mind, we got another useful cheque today, — this month's statement should look a lot better.

Tuesday October 6th

Roger arrives and we crawl into the wheat. He's got lifters on the combine, and a new razor-sharp knife, — but it's still a battle. The straw builds up at the front of the table, bulldozing, — the elevators block up, the drum's full of soil, and Roger's in a moderate fettle. It takes ages to get a load into the trailer, and then I get stuck leading off just short of the gateway. We have to get the other tractor and tow the first one out (more delay). After that it's half-loads only.

Then while I'm tipping in the shed, the hydraulic pipe on the trailer ram blows, — oil everywhere. Rush to borrow Charlie's trailer (more delay). Roger's not amused. He has to be to Arthur's tomorrow (Arthur won't be amused either).

We struggle on until ten o'clock at night, by which time we've got the last seven acres surrounded.

Wednesday October 7th

Hailstones. Would you believe it? — hailstones the size of pullets' eggs!

Thankfully it doesn't last very long, — but no harvesting this morning. Ye gods, anybody who can make a living farming in this climate can surely do it anywhere. But maybe it wouldn't be quite so exciting.

The wagon comes to take my contribution to the mart. I see them graded and hurry back in case Roger decides to start again after dinner. He does. He's still got two hundred acres to cut in this parish, — and he's keen to get going.

The moisture content is up a few points, but we begin again at two o'clock, and eventually scratch through the seven acres in four hours. He says he'll have to charge a bit extra for this field. What can I say?

The harvest is finished. It took two months.

Thursday October 8th

We have a heap of scruffy wheat in the shed, — but at least we've got it! Since Hurrican Harriet the price has gone up a couple of quid, and may go further, — but I'm in no position to play the market. I do a quick deal on the understanding that it will be paid for by the end of the month, and moved by Monday.

Charlie's still got a field to cut (he's next but one on Roger's list) so we both went to a store cattle sale in search of 'wintering' material. The trade was 'buoyant' (that's the word they'll use in the papers tomorrow), — crackers might be more appropriate. Charlie seems undismayed, and buys a pen of big rough bullocks early on. Perhaps I'm influenced by this irresponsibility, because I come away with a miscellaneous selection of smaller young stirks which will hopefully grow into something useful by next spring, — but Charlie's lot look the better value (not that I'd ever tell him).

Friday October 9th

Harry, the hairy hill farmer, drives in eager to buy the last field of wheat straw. I give him a cup of tea, and listen patiently to his pathetic tale of the deprived and desolate existence of those who struggle in hill-cow country. You would imagine, to hear him talk, that he lived with the yeti half way up the Himalayas. Well, fair enough, his wife's a formidable woman, and they do have a genuine hill farm, — but they take no harm. As Percy says, 'those buggers get subsidised every time they break wind. . . . ' For a lot of the year they do nowt, lambin', calvin', clippin' and dippin', — that's it. No rape blowing out, no flat wheat, no mildewed barley, . . . wait a minute, I'm beginning to talk like *him*!

A cruel but reassuring little story in the pub. Seems Cecil sold some milling wheat at an enormous premium (there's a big demand for good clean wheat that passes all the tests, especially since the hurricane) and, as you would expect, Cecil's did. But of course you can't count your milling premium till it's in the bank (until it's bread). Sure enough the stuff was rejected at a mill near Glasgow, and poor old Cecil had to pay the haulage back to a feed store in Yorkshire, — and then was paid the same as we lesser mortals.

We lesser mortals had a drink on that.

Saturday October 10th

Harry arrives with his two monster sons, both built like brick netties, and begins to bale the wheat straw into big round bales (they'll probably carry them home, one under each arm).

Walter's man is spreading lime on the ploughing, and spraying the weeds in the rape.

Willie is recovered, and playing rugby in some unexplored region of County Durham. Gladys is Christmas shopping!

Get the lambs in, and shed off about eighty that won't get fat off the grass. They will go onto the turnips together with some more bought-in stores. This leaves forty to sell.

'Hot-Rod' comes to take Doreen out, — mind she does look smart, I fear this could be serious.

Sunday October 11th

Hairy Harry and his 'delicate' sons must've been here most of the night, — they've almost got the field cleared.

It's time we had the wheat sown, — even Davy Scott has a field in. I phone Roger, who says he's on top of the combining now and can send a plough in tomorrow, followed by the drill early in the week. All we need now is the seed.

Monday October 12th

One of Roger's machines is ploughing the wheat stubble, — and making a canny job. It's not easy 'cos there are still patches of flat wheat we couldn't get, and a lot of long uncut straw. My wee three-furrow friend would be hard pressed to bury all that trash.

Make a desperate phone call for seed. It's ridiculous, — these are the same people who threaten me with some kind of penalty if I take five minutes too long to load a grain wagon at harvest time, and now they're a week late delivering seed. I am not amused. Of course the silly little secretary, carefully trained to make coffee and excuses, says she's sorry, but it's not her fault. No, I suppose it isn't.

Tuesday October 13th

Cultivating land. Where do all the stones come from? Every year I pick tons off this field, and every year some nasty rock-making devil down below pushes some more up to the surface.

I finish working one field, as Roger's bloke completes the ploughing in the other. I was leading boulders off when the seed lorry arrived. The driver obviously expected me to be sitting there waiting for him and had the audacity to tell me he was behind schedule.

By the end of the day one field is ready to sow and the other disced twice, and I have stone-picker's back.

Wednesday October 14th

Roger's drill arrives at breakfast time, — panic! Run about like a headless chicken loading trailers, then away to cultivate the second field. He'll catch me up, — there's no time for dinner. Load trailers again. Thank the lord for pallets and the forklift.

Roger's man is one of those Rambo-type youths with the strength of an ox. Never tires, only has one gear (top), eats standing up, throws bags of seed about as if they were pillows. A slash of the knife, fill the drill, and away down the field again.

Suddenly it looks as if we'll be short of seed (more panic). Rush to Arthur's and luckily he has a bag left over. It's a different variety, but who cares. Another bag from Fletcher, and that sees us through. At nine o'clock the drill is back on the transporter, and the winter seed is all in the ground.

Thursday October 15th

There's an air of anti-climax, a tendency to do nothing, a sneaky little voice saying, 'right bonny lad, you're finished, y' can take it easy now ... maybe.'

Walter Williamson comes to arrange some pre-emergent spraying and is overjoyed to tell me the slugs are rampant this year, and I should have pellets on everything, — now!

Michael McMurdie phones to tell me he has a store lamb sale tomorrow, and he needs some more entries. I hope he gets them, 'cos we need a few for the turnips.

Hairy Harry also phones (didn't think he'd have a phone) and offers to sell me some stirks. I might go and have a look at them.

McMurdie phones again asking for entries for the 'fat' on Wednesday as well.

There's a lame tup. We've got to get him fit, he's got some important work to do soon.

A rep calls to sell nitrogen. Delivery now, pay next year.

The hedges need cutting; there's a burst drain; the bullocks are systematically destroying the fence between them and the rape. No, we ain't finished yet.

Friday October 16th

It's a frosty morning, — a good chance to spread slug pellets, without making a mess. However by the time I hitch up the spinner the rain has started, ('Ah well, that's life,' he said philosophically ... in fact what he probably said wasn't very philosophical at all, — but there wasn't anybody else about, — so it didn't matter.)

By dinner time I find myself at the mart having bought fifty store lambs.

Cleaned out the dipper, had a bath and went to the pub. Here there was the usual mixture of agricultural customers, some of whom were very pleased with themselves, having finished harvest and autumn sowing, and others who were very quiet, 'cos they hadn't. Davy Scott still has ten acres of oats to combine, — but he'll end up selling them for a fortune to Knickers and Co. for their horses.

Willie has 'broken up' for half-term. He doesn't know it yet, but he's dipping sheep tomorrow.

Saturday October 17th

Dipping. The water is almost freezing. Some of our more mature ladies wouldn't survive for more than ten seconds in the tub, and they know it. The stubborn auld buggers resist all attempts to push them towards the drop, digging their toes into the slightest crack in the cement. It's a battle with every yow. Occasionally one will fool you by doing the running jump trick, and only falls from flight onto another unfortunate yow

below, who is paddling gamely for the steps and survival.

There's a lot of colourful language. Gordon has a vast vocabulary, and I think Willie's is substantially enlarged by the end of the operation.

We have them all through by one o'clock. Gordon and I are cold, wet and miserable, and I expect the sheep are too, but Willie swallows a bacon sandwich, and hurries off to play rugby. If he handles the opposition like he handles male yowes, he could be sent off today.

Sunday October 18th

Fletcher, Arthur, m'self and Rodney the solicitor play golf on a bright, crisp, quiet morning. A perfect autumn day, — the grass is still green, the leaves are still on the trees (even though, any day now, a frost and a breeze will remove them). Mother nature is perhaps giving a gentle smile and an end-of-year cuddle, just to persuade us that she's not always a bad-tempered bitch.

We the players match her gentle mood all the way round. The golf is chatty, relaxed, good shots congratulated, bad ones ignored, — we laugh at silly jokes, give a friendly wave to those on another fairway. Eighteen-inch puts are generously given, and quickly taken. All very English, very civilised, very Sunday morning.

Until the fourteenth tee.

Here it was that Arthur came to grief. He and I were three down at this stage, but still quietly confident of saving the game and the drinks. Everybody else had driven off, — Rodney was out of sight, I was out of bounds, and Fletcher was in the rough. There was a bit of pressure on Arthur to get a good 'n' away. Well, he's a strong lad, and if he really connects it can be awesome. So he lined himself up, took a couple of practice swings, muttered a few threats, and launched himself into the drive. Head still, slowly back, just a suspicion of pause and wobble at the top of the swing, and

down comes the club to swipe the innocent little white thing into oblivion.

Perhaps he topped it, perhaps he got underneath it, perhaps he hooked it, sliced it ... we'll never know, because the ball disappeared straight up his left trouser leg!

Arthur was surprised and disappointed. In fact he should've been very relieved, because it didn't get beyond his knee. We lost 4 and 2.

Monday October 19th

The lambs I bought on Friday are of the 'homing' variety. They're rakers. No matter where I put the sods, they get out and immediately head in a westerly direction, − consuming everything in their path like a plague of locusts. Already they've sampled the rape and the barley (which is just peeping through, and must've been amazed to find itself face to face with a ravenous cross-Suffolk hogg). They've been up the road into Charlie's barley as well, and into the village twice, eating the Pillicks' lawn and a few late-flowering shrubs (with a bit of luck we won't be invited back there). They're remarkable beasts. They can graze and run at the same time, which has to be bad for the digestion. They'll probably never get fat.

This morning they're back at the mart scratching about in the car park! I get them home before McMurdie sells them *again*.

Tuesday October 20th

Willie and I go up to Harry's place to view store bullocks. I've been there before, maybe ten years ago. Nothing's changed, − it's a mediaeval menagerie. His collection of mechanical things is even older than mine, so much so that they're undoubtedly worth a lot more. Miscellaneous species of livestock wander unhindered by fences. On the track up to the farm we have to negotiate our way around dozing cows and calves, all known breeds, multi-coloured, horned, poled, all ages, ancient and modern, − and a bored looking horse. Blackies and Swaledale sheep are roaming about everywhere, though the two yowes lying at the yard gate obviously stopped roaming several days ago. In the yard ducks, hens and geese cackle among the clarts, a collie bitch dozes by the back door, her litter of black-and-white pups asleep and interwoven into one warm bundle, inside a capsized bucket.

The house cow peers out of the byre and bellows some kind of greeting. A dog somewhere else begins to bark. The geese join in, and Harry and his sons emerge from a seventeenth-century stable, and tell everything to shut up. And they do.

The stirks are in a low dark building, made even lower and darker by the rising tide of maybe three years' muck. In the gloom the animals look enormous, and I ask to see them outside in daylight, and on hard ground. This is easier said than done. For a start, we have to lift the gate off its hinges, and then the cattle are reluctant to venture out and down into the unknown. We spend fully half and hour running round and round this hemmel, hitting our heads on the beams, and screaming abuse at the bewildered beasts, until at last one or two jump out. The rest soon follow, except the inevitable *one* of course, — but eventually even he takes the plunge.

The problem then is to keep them in the yard. They're mad, — several try to climb the wall. The lads position themselves at the gates, armed with forks, defying anything to pass. Harry hoys a bale of hay amongst the cattle, and finally they settle down. Only then can I get a proper look at them, and negotiations begin.

There are twenty-five in all; twenty will do nicely. They're good sorts, decent back ends, good bone, nice heads, canny beasts.

We barter back and forward. I say they're not very big, he says they'll grow like mushrooms. I point out a poorer one, he indicates the best 'n'. I ask what he wants for them, he says make 'im an offer.

It takes us half an hour of sparring before we get to the real money and, as expected, he wants a tenner too much. 'Right,' says I, 'I'll give you the price, but you keep the smallest five, I just want twenty. ... ' Before I've finished the sentence he's shaking hands. It's a deal. God he was quick at the finish, — have I paid too much? 'And you deliver them,' I say, while I've still got hold of his hand.

Wednesday October 21st

Drier conditions so I spread slug pellets early morning. Already there are stories of patchy crops. Even Cecil has a thin field of barley, — and if slugs have the audacity to attack him, nobody's safe!

Sent six bullocks to the mart, the last off the grass I think, and a few lambs (not very fit, but that seems to be what they want nowadays).

A fair trade, but the cattle didn't weigh any better than expected, — and Ghadaffi took a long time to grade one of them.

Thursday October 22nd

The homing lambs have settled down, and are mixed in with what's left of our own. They can all go onto turnips soon.

Get the yowes in, push them through the footbath and divide them into four lots, ready for their once-a-year orgy. The lame tup is more or less sound, so hopefully all the 'caballeros' are working themselves up into a frenzy.

Friday October 23rd

Bought half a dozen moderate bullocks at the mart, — definitely paid over the odds for them, and they look no better at home. Willie and I wormed them after tea, and left them inside for the night.

Charlie bought a single wild-eyed cross-Friesian bullock that galloped round the ring with its tongue hanging out. He got it very cheap.

Finished the slug pellets just before the rain and darkness. It's a filthy night.

Saturday October 24th

It's my birthday. I get three cards, a crunchie bar, some after-shave and a pair of reinforced wellie socks. (It's Gladys's next month, don't forget.)

Jesus, the years flash by now, — seems the older you get the quicker time goes. I don't remember exactly when it was, some mid-life crisis I expect, — a bad lambing maybe, trying to catch a gimmer on a wet night, and breaking down knackered ... but quite suddenly mortality stares you in the face, and you wish you were just a little younger. Later, like most old men I'll probably pretend to be older than I really am. Gordon is always telling people he's nearly seventy (as if it was some Olympian achievement). Actually he's sixty-six. As soon as he *is* seventy, he'll say, 'gettin' on for eighty now y' know. ... '

Charlie and I were once discussing old age. Not in any morbid way you understand, in fact we were quite delighted to have got this far, having managed to avoid killer cows, radioactive sheep and cowping tractors. As Florrie says, 'there's not one in a thousand gets out of this life alive' ... and history would suggest that even those odds are somewhat optimistic.

Anyway it was a few years ago now when I realised the full flush of youth had passed, when even very old yowes, with feet like Chinese mandarins, could leave me with just a handful of wool as they sidestepped past. When policemen suddenly appeared as spotty twelve-year-olds in big boots, — when pretty girls in summer frocks stopped smiling back at the dirty old peasant who leered at them. When reps became experts in their teens, and my own kids abandoned the fragile in-bred pretence of parental respect, and replaced it with blatant incredulity.

And now the equipment's going, — the knees, the wind, and above all the memory, — in particular people's names. One day I bumped into a bloke in Bridge Street. I knew him well enough, I was sure of that, — but who was he? Well, he had to be a farmer, — I only know farmers. ...

'Hiya,' says I cheerily (a lot of people are called Hiya, when you've got an affliction like this) ... 'so how's the lambin' gone this time?'

'The lambing?' he says, 'what bloody lambing?'

Of course I should've just run away, or thrown m'self under

a bus, — but I didn't, I blundered on. 'Oh, don't you keep breeding sheep any more?' I asked. 'Look,' he said, 'I don't know what you're talking about, — but I work in the bank. I cashed a cheque for you twenty minutes ago. ... '

Well, I knew I'd seen him somewhere. I wonder what I did with the money?

Sunday October 25th

The clocks went back an hour last night. We woke up at 6.30 as usual, realised it was only 5.30 and slept in till 8.30. Why the hell can't they leave the ruddy clocks alone?

The last of the townie bramble pickers are out in force. You might imagine that just having left 125 Mafeking Gardens they'd want to park in some secluded private spot. But no, as soon as they see another car parked in a gateway, they pull in, — mother and the kids scratch about in the hedge with a tupperware box, Dad slackens his braces and reads the paper. Within minutes it's like a car park at the Supermarket. Funny how townies have the herd instinct, and we rural creatures prefer two hundred acres to ourselves.

Monday October 26th

This is it lads, your big day, — the festival of erotica that will leave you but a shadow of your former selves. I harness up the tups and take them out to meet the harem, — top lips a-curlin', they know what's expected of them.

I recall when Willie was only about six or seven, it must've been this time of year. The tups were in the hemmel, harnessed for action, and Gladys sent him to feed them with some cabbage leaves or the remains of a turnip.

He came back with a puzzled expression. Something was obviously bothering him, and he eventually asked what the harness and the crayon were for. Gladys figured this might

201

be as good a time as any to tell him about the facts of life, — she might not get a better chance.

By the time she'd finished her lesson on procreation, Willie had an evil grin on his cheeky little face. When I came in for m' dinner the little bugger asked me where I kept *my* harness!

Tuesday October 27th

Sleet, rain, wind, — a cold moderate back-end day. It must be difficult for a tup to work up much enthusiasm in such conditions. Each of them has a few marked, but this morning all the sheep are up against the dyke back, looking far from amorous.

And this is the day Harry decides to deliver the cattle I bought last week. They tumble out of the wagon, and look for a wall to jump over. They'll have to go inside, we'll never keep them in a field.

Harry gets his cheque, and I get a few quid back in cash for luck. I think they'll be alright, they're not pretty, but there's potential there.

What a gamble farming can be, — you lay your bets and it could be a year before the 'horse' comes in. And of course, it might fall.

Wednesday October 28th

Davy Scott finished his harvest at the weekend, so it's probably safe to say that all the farmers in the parish have
everything gathered in. The vicar should never have the
Harvest Festival until Davy gives him the nod.

Davy's a canny bloke. It's surprising he's never been married, — because (as already indicated) he's quite partial t'
women, — and several women have fancied their chances with
him, but nothing serious has ever developed. However, I *do*
remember one bad winter when he got very fed up with
looking after himself, and advertised for a mate in *Exchange
and Mart*.

'Destitute peasant requires female companion housekeeper,'
it read, — 'applicants should be proficient at clipping sheep,
singling turnips and rodding drains . . .'

But Davy reckoned none of the replies were suitable.

Hitched up the plough and started on the front field. This
is for spring barley next year. No hurry, but it would be good
to get the land turned over and let the frost at it.

Thursday October 29th

One of those days.

Looked at the stock and found a lamb stone dead. It had
given no warning, obviously wanted to surprise me.

One of the tups has pneumonia I think, — give him a jab
and hope for the best (but I might have to hire a replacement).

Another has apparently just stopped performing, — no
yowes marked for two days. Have they all got headaches?

Harry's stirks have eaten the hemmel gate and discovered
the hayshed. They've eaten their way into a ruddy great hole,
and the rest of the bales are threatening to collapse on top
of them.

One of our own bullocks has gone to chat up a bulling heifer
at Charlie's, via the winter barley. Takes me all morning to

get the fool back and mend the fence.

Ploughing after dinner, — the tractor dies! I forgot to put diesel into the beast, didn't I. Walk home for fuel, and bleed the system.

Just get going again when a wagon comes in with fertiliser. I'm removing the third pallet when my foot slips off the clutch and a lot of bags end up on the floor, — burst. Then we get a puncture in a back tyre, and half the load has to be taken off by hand. Phone the tyre company at tea time, — the phone's off.

Gladys at WI meeting, leaves supper in oven to heat up by eight o'clock. I fall asleep while watching rubbish on television, and by nine the supper is incinerated. Eat a marmalade sandwich, fruit cake and biscuits, with four cans of beer.

Willie and Doreen miss the last bus home from the pictures, and get a taxi. Surprise, surprise, — they haven't any money to pay the driver.

In bed at midnight. Forgot to switch the blanket on!

Friday October 30th

Consider spending the rest of my life in bed where it might be safer and easier. However, the four beers insist I get up at 6.30 as usual, and begin to repair yesterday's disasters.

The phone is still dead, Gladys by now in a highly nervous state being unable to contact Boadicea or Hilda to discuss issues of worldwide significance, such as the price of curtain material.

I transport the dead lamb to the kennels, and call in at the Tyre Company to arrange for repairs.

Home to inject sniffling tup. Reinforce hemmel gate with battens and barbed wire (it's like Colditz now), prop up the leaning hay bales, have m' supper, and go to the pub.

Gladys is happily blathering on her reconnected life-support machine.

Charlie's wild-eyed bullock (the one he got cheap last Friday) has escaped and redesigned the Pillicks' garden. The poor Pillicks must be a little disillusioned with rural peace and tranquillity by now. They'd probably be better off in Brixton.

Arthur has had a riot among his potato-picking gang. These happy little helpers are a battalion of militant housewives, all of whom look like Les Dawson, and are organised by a 'shop steward' whose word is law. Those of us who have ever employed this type of temporary labour know that to fall out with them can be dangerous, — they take no prisoners!

Arthur it seems made two mistakes. He failed to measure their individual lengths to the nearest millimetre, and one old witch discovered she had almost a foot longer strip to pick than anyone else. Secondly he negotiated a rate of pay per day rather than per hour worked. Then when the heavens opened and the pickin' was abandoned at dinner time, the evil old crows demanded full wage and a bag of spuds each.

When Arthur protested he was lucky to escape with his life. He still has three acres to harvest, — by himself.

Gordon says it's going to be a hard winter, — something to do with an abundance of rose hips I think, or was it the

worms on his lawn, or the hedgehog sleeping in his coal house,
. . . I can't remember. I do know that McCaskill is forecasting
a hard frost tonight.

Saturday October 31st

Sure enough, a very cold frosty day. The grass is about
finished in this part of the world, and the bullocks will have
to come inside soon. I put the electric fence onto the turnips,
ready for the hoggs, and go watch Willie play rugby.

He had a useful game against some High School from
Durham, scored another quick 'ferrety' try from five yards
out, missed a kick under the posts, and punched a wing
forward twice his size. Our lot won 24 — 12.

There's a disco tonight after the match, so I'll have the little
waster to collect at some ungodly hour again.

Roderick and Doreen are out to dinner. Doreen in a very
revealing frock which may put young Rod under considerable
pressure. They're celebrating their anniversary, — been
together now for five whole weeks!

Gladys and I (who have been together a little longer than
that) fall asleep while watching Match of the Day.

I go for Willie at midnight, and find him lying mutilated
in the car park with the ever-lovin' Tracey bathing his wounds.
Apparently it is a case of the wing-forward's revenge. Serves
'im right!

November

Sunday November 1st

Willie stays in bed all morning, groaning pathetically, but demonstrates his dramatic powers of recovery when the aroma of roast beef and Yorkshire pudding wafts upstairs.

Doreen is 'lunching with Roderick's people' (would you believe?)

Gladys persuades me to take her for one of her 'runs in the country', — so we have a tour round by Cecil's (disappointing, no disasters there), up to Arthur's (his barley's looking good), on to Fletcher's (he's got a miss in a field of rape, — half the drill empty for about twenty yards, I suspect).

Back into suburbia to have a sneaky look at Roderick's 'stately home'. It's nowt flash really, a semi in Acacia Avenue, but Gladys thinks it's a 'nice area' and they must be nice people. I tell her Crippen lived in a house like that, but she's not listening.

Monday November 2nd

Sweep and I bring the hoggs into the pens, dose them (again), put them through the footbath (again) and move them onto the turnips. They only have a few drills to start with, till they get a taste for their new diet, but they fancy the leaves, and it's eyes down for a full bite.

I stay with them for a while, until they find the electric fence, have a sniff at it, get a shock and jump back. (Sweep knows all about electricity, and stays at the gate.) They look good, but if the weather breaks, they might need a run off onto a dry lie.

The tups have been out for a week now, and appear to be working alright, about a third of the 'girls' have blue bums.

Roger's machine is tidying up the road-side hedges.

Tuesday November 3rd

Back to the plough, the endless furrow and Radio Four. — But it's hardly the relaxing job it can be, — because now the land is wet, the plough has to be 'eased' through the soft spots to maintain a reasonable depth. I remember the first time I ever ploughed, — on an Old Fordson with spud wheels and a Ransome trailed behind. No cab, no heater, no radio, — wrapped up in balaclava, duffle coat and woolly gloves, my instructor sitting freezing on the mudguard. I was leaning back adjusting depth, width or whatever, anxious to impress, leaving the tractor to find its own way up the open furrow, when the headland came unexpectedly. The tractor continued into the hedge and climbed halfway up a plum tree. The instructor fell off into the clarts.

The plum tree is still there, none the worse really, but every time I pass I apologise to it.

Wednesday November 4th

A foul, cold, wet autumn day, trying hard to be a winter's day.

The hoggs are beginning to nibble at the turnips. Inevitably one of them has discovered that the hurdle at the end of the electric fence isn't electrified, and has jumped over. Of course he can't find his way back, even when the hurdle is opened. After running up and down for a while, he eventually dives into the net, wraps himself up in it, and is caught. By now I'm blowing and spitting and cursing, so he gets a bloody good hidin', and is hurled back amongst his mates. A length of barbed wire stretched along the top of the hurdle, *might* keep the pest at home.

The bullocks are slipping, — they're developing hairy winter coats, one or two have runny noses. It's time they were inside.

Bed the hemmel, clean out the water trough, fill the hecks with hay. It looks warm and inviting, I could almost spend the winter in there m'self. By the darkening the bullocks are lined up under cover chewing nicely. Some of them won't take a lot of finishing, others will be there until next spring. We've had one of them a year already, he's gone badly. He's a bit like Arthur Thompson, — no matter what he eats, he doesn't put any flesh on. Maybe he should've been sold in the store for somebody else to stuff (the bullock I mean, not Arthur. ...)

Thursday November 5th

Changed the markers on the tups. The belly's certainly gone off them, but what can you expect after days of unrelenting passion, — and they ain't finished yet.

That lamb's over the net again, sooner or later some of his pals will follow him. I have two alternatives, — shoot the sod, or send him to the mart next week. We'll try the mart first.

Doreen and Willie go to a bonfire party at the rugby club. It's washed out twenty minutes after the fire is lit.

Friday November 6th

The winter has begun. I hate November, it can last until March. Autumn is gone, all we'll get from now on is sleet, snow and cold winds. Oh yes, we might get the odd canny day, but for a long time Mother Nature will be in a moderate mood.

Meanwhile back on the farm all is not well. An army of

slugs is marching over the wheat (we've pelleted them once already, but they're back again). And a vast blanket of pigeons and crows has descended upon the rape.

I borrow Charlie's trike and begin a massive counter-attack, spending all day bouncing across fields blasting everything that moves. I think I might've shot a slug, — but I couldn't find the body.

Left the banger on all night, and took the phone off the hook.

Saturday November 7th

Slugs, crows, pigeons, — and now the blasted hunt again!
They met at Cecil's, — the slimey toad is always trying

to curry favour with the upper crust. Even Wayne (the creep) has got himself a horse and joined the cavalry. I suppose he figures that the aristocracy might offer him another farm. (If you want to get ahead, get a horse!)

By the time they get to our place they're all 'tanked up'. Cecil has been generous with the pop, and several red-faced right honourables look decidedly insecure as Knickers and his men draw the turnips.

Now I've mentioned it before, — this bird scarer of ours is a fickle piece of equipment. Sometimes it just clicks for an hour or two, then it pulls itself together and begins to blast away every five seconds. It can be like the barrage at El Alamein.

Well as you know, I'd left it on all night, and I think it must've been in a clicking mood most of the time, — but at exactly twelve noon it got its act together. It wasn't as close to the horses as last time, but it was close enough to upset them.

They soon moved on (in some disarray) towards Charlie's place.

Maybe if I put a bird scarer in every field the hunt would bypass me.

Sunday November 8th

Boadicea came for dinner. She's one of those indestructible spherical old bats, always immaculately turned out, eats and drinks as much as four long-distance wagon drivers, never has a cold, and wishes Gladys had married an accountant or a solicitor, — somebody with a suit. Sweep always jumps up and covers her with clarts when she arrives. I think he does it on purpose.

Roderick came as well so there were six of us at the table. The beef was superb. Nothing beats a good bit of beef, — I could eat it every day. But sadly on this occasion there'll be nowt left to have cold tomorrow. Willie, Roderick and Boadicea see to that!

They all went for a walk after the washing up, while I read the Sunday papers. There was an article in one of them forecasting more gloom and ruin for farming, written by some whizzkid analyst in Brussels, who'll be on thirty thousand a year, and a ruddy great pension. I wonder what he's doing this afternoon? I have a kip.

Woken up at tea time by four cyclists all with punctures. They're blaming thorns on the road where Roger was hedge cutting.

Monday November 9th

Charlie has had a fire in some outbuildings, — faulty wiring they reckon.

Apparently he was standing at the kitchen window having his first cup of tea this morning when he noticed a glow behind the stable. It took him a few minutes to realise it was in the wrong place for an early sunrise. By the time the penny dropped and he'd leapt into his wellies, the old barn was ablaze.

Give 'im credit he didn't panic. (I mean he didn't phone the fire brigade or anything silly like that.) He had a careful think about what was in there, about what else might be at risk, then spent half an hour moving some calves away from the smoke before he got Hilda up, and asked *her* to phone the brigade. Long before they arrived he'd managed to lose a lot of very poor hay, most of his antiquated sundry farm tools (ladders, shovels, a wheelbarrow, — that sort of stuff) together with the little old Massey tractor he was going to replace anyway (it wouldn't start so they had to push it into the barn).

He reckons it was a 'useful' fire.

Tuesday November 10th

Spent the morning at Charlie's, helping to tidy up after the inferno. Newcassel-Browne was there assessing the cost of a new roof. Knickers had trotted across to view the damage from his usual vantage point sixteen hands above ground level.

Charlie will have to buy some hay (he'll want a lot better quality than was burnt). He's already been down to the store and re-equipped himself with shiny new shovels, brooms with bristles, a wheelbarrow complete with a wheel, two steel ladders with a full complement of rungs. A tractor salesman arrives as I leave. No trade-in; he should get a good deal. I go home to sit on my old Ford and finish the ploughing, — all except the headland, — on a cold drizzly afternoon.

Wednesday November 11th

Houdini wasn't over the net this morning. Did he know of my plan to send him to the mart? Have we a mole, an informer? Anyway I couldn't pick him out. I'm not like Arthur Thompson who can tell you where he bought every animal he's got, which farm it came from, how much it cost, what medication it might've had. If it's a yow he'll tell you (whether you want to know or not) that she had twins last year, a single the year before that, and triplets as a gimmer, one of which died of watery mouth on April Fool's day.

So our leaping, wandering nuisance survives, hidden for a while.

I went to the mart anyway, just to see the trade, and have m' bait with Charlie. Gladys *expects* me to be missing on a Wednesday, — I'd be lucky to get fed if I stayed at home.

Thursday November 12th

A difficult day, — appointment with accountant.

This is a jacket and tie occasion, clean shoes, change of socks. I'm never sure what we're going to talk about, — but I go prepared with livestock numbers, bank statements and a few other crucial figures scribbled on an empty fag packet. I've seen Cecil go in with a briefcase. I even thought about getting one m'self, — they look very impressive. But it seems silly to have a posh briefcase, just to carry an empty fag packet.

The accountant always asks a lot of awkward questions about assets and liabilities, how many bales of hay or turnips I've sold for cash, how much housekeepin' Gladys gets, and so on. I figure most of this is none of his business.

He gets excited when I tell him how much fertiliser we've got in the shed, and how many acres of winter corn's sown, — until I tell 'im none of it's paid for yet.

He comes to the conclusion we're on the brink of bankruptcy, which is quite reasurring. He told me that twenty years ago, so we must be holding our own!

Personally, I feel we're in a fairly sound position, — fully stocked, plenty winter feed, all the corn sown, still a bit money due for wheat, and the hoggs to sell. Doreen's saving us a few quid every week now, and Montague hasn't bothered me for ages.

It's good to get home and put the working clothes back on, — I've been freezing all day in the posh gear.

That bloody lamb's out again, — I may kill it ... if I can catch it!

Friday November 13th

A foul bleak typical November day, — rain with a dash of sleet coming in off the North Sea. The yowes (who aren't very bright) are huddled together in the most exposed spot they can find. The hoggs on the turnips are all lying on the headrig (except one, guess who?)

I can see Charlie's cows standing (humpy-backed) at the gate, waiting for their next feed. Surely even the slugs are sheltering today.

Florrie is vacuuming, dressed seductively in four jumpers, a pair of ex-army trousers, and a beret pulled down over her ears. She says she's got 'Austrian arthritis', and all her joints seize up in the cold weather.

Saturday November 14th

Another desperate day. The countryside is awash, little lakes forming all over the place, and still the wetness blows in. The spuggies and starlings are sheltering in the hemmels again. The tups haven't performed for two days, and you can hardly blame them. Let's face it, the prospect of sex with a wet, toothless five-crop mule in the middle of a twenty-acre field must be fairly daunting on a day like this.

It's Gladys's birthday tomorrow. I have secretly purchased an exotic potted plant, a box of fudge, a space-age tin opener, and a card covered in hearts and flowers. I bet she'll be chuffed t' bits.

Sunday November 15th

I had anticipated a touching little scene at breakfast time when I handed over my presents, and sure enough she was quite surprised. Her birthday's *next* Sunday!

Meanwhile it's still raining. There can't be any water left up there, — the turnip break is a bog that sucks your wellies off.

Charlie says his cows are making a hellova mess, plunging in over the hocks, and always blarin' for more grub. I've seen him feeding them. He sets the tractor off in bottom gear, jumps out of the cab, scrambles onto the trailer, throws the silage off into heaps as the tractor meanders down the field, then climbs aboard again and roars for the gate. He just makes it. The cows can consume everything within minutes, and if he's not very quick, they'll have him as well. I recall they

nearly *did* catch him once, and ate all the electric wiring on the Ford 5000, and most of Charlie's coat. He was lucky to escape.

McCaskill says the rain and sleet will ease off tomorrow, then we'll get an unsettled spell. Great!

Monday November 16th

Only thirty-four shopping days to Christmas. I don't really wish to know this, but Gladys informs me as she sets off on a spending spree. 'Can't leave it all till the last minute,' she says, 'and anyway if everybody was as miserable as you, — there'd be no Christmas at all. ... '

Would this be a tragedy? I ask m'self.

Meanwhile back in the real world, I bed the cattle. Harry's stirks have settled down, — they no longer climb up the wall when I go in amongst them and, having seen the climate outside, they haven't tried to escape again either.

The hoggs are in a mess. They've got nowhere to lie down poor things. One emaciated runt has lain down in the clarts, and may not get up again. Houdini is over the net and is busy taking one bite out of every plant in the field. He's gotta go!

McCaskill was right (more or less) it's just drizzling by nightfall.

Tuesday November 17th

I caught 'im this morning (the hogg that is, not McCaskill). He got himself wrapped up in some brambles, so I grabbed the little rubbish, dragged him home and locked him in the byre until tomorrow.

Shifted the nets on the turnips, put a couple of hay hecks in, and left the gate open for the sheep to run back into a grass field. Carried the runt onto the grass, — at least he can put one foot past another there.

Gladys has a mountain of brightly coloured boxes on top of the wardrobe. I can't find anything with my name on it.

Wednesday November 18th

To the mart with the hurdling hogg. Even with his legs tied he managed to scraffle out the back of the pickup, so we went together in the front, — me driving, him on the passenger seat. I swear he waved to Mrs Simpson, as we went through the village.

Ghadaffi rejected him of course, and to be honest I feared as much, — he was pretty lean really (rakers often are) but I sold the wretched thing on a half-weight basis anyway. I only found out later that Humphrey Smith was buying all the ungraded hoggs for Cecil, so the great Houdini escapes death again!

Thursday November 19th

Gladys is having a coffee morning to raise cash for battered kids, or battered wives, — maybe both, — anyway the house is full of big-bosomed women blatherin' on about frocks and Christmas cakes. Charlie and I decide to seek refuge at a mart somewhere (anywhere!)

We eventually end up at a store sale at a mart neither of us has been to before. We hadn't really intended to buy anything, just have a look, a pie and a pint, and home before dark. However there was hardly anybody there, and the trade was quiet, and we couldn't resist a flutter.

Charlie started it, bought a good in-calf heifer for next t' nowt, and after that we thought we might was well fill a wagon. I ended up with ten rough bullocks, and thirty little cross hoggs worth the money.

Friday November 20th

Yesterday's purchases didn't arrive until one o'clock this morning. The driver got lost. Said he'd never been this far north before, and wouldn't have found us at all if it hadn't been for someone in the village. It seems he'd seen a light on, and knocked at the door to ask directions. This lady had answered, dressed in not very much, and insisted the poor lost soul had a cup of tea (at least) before he went any further. Sounds like Mrs Simpson.

The new hoggs can go onto the turnips in a few days. I'll give them a pulpy kidney shot first. The bullocks have never been inside, and they'll have to stay out now, — there's no room for them in the buildings anyway. They can go down by the wood, it's the driest bit of ground we've got.

Went to the Swan after supper. Charlie's chuffed to bits with what he bought yesterday, — four in-calvers, and one old cow with a good bull calf.

Cecil has a lamb missing, — I know who that'll be.

Saturday November 21st

The emaciated runt has expired. He never moved from the nice dry spot I carried him to on Tuesday.

The tups have completed their nuptials I think, but we'll swop them around, and leave them with the ladies for a while yet.

Willie played rugby in a sea of mud. After quarter of an hour only the ref was recognisable. They drew six apiece. When he came home Gladys made him have another bath. He said he'd had one already, — but there was still a lot of clart in his hair and lugs.

The hunt didn't bother me today, but rumour has it they chased Arthur's hoggs through the nets on his turnips. Arthur is not amused, and has threatened to shoot the next horse that sets foot on his farm.

Sunday November 22nd

It is Gladys's birthday, and she gets lots of goodies. Unfortunately my exotic plant has died, the fudge is long gone, the space-age tin opener has disintegrated, and my romantic card has been put away in a drawer. Apparently it was a 'Get Well Soon' card.

Charlie and I have decided to go to Smithfield next month. We've never been, but everybody else has, it seems. Even Harry the hairy hill farmer once went to the show with a pair of Blackface ewe lambs in 1978. I remember he was arrested in some seedy night club. He'd refused to pay the bill, and a cockney bouncer had tried to throw him out. Of course after a lifetime spent with wild Blackface yowes and cross-Galloway stirks, the bouncer was no problem. It took most of the Soho Mafia to get Harry onto the street. They took all his money though, and he's never been beyond Gateshead since.

Monday November 23rd

Montague grabbed me in the bank this morning, said he wanted a quick chat in his office, but I persuaded him I was in a tremendous rush to get back to a sick bullock. He said he'd phone me later.

When I got home Gladys took some more cash, and went off to Marks and Sparks for more Christmas goodies.

On the turnips, the battery on the electric fence is flat. The hoggs haven't realised it yet, and once they've had a few shocks, they might never touch it again, but we're not going to risk it. I find a spare tractor battery and connect that. It'll do just now. Gives a hellova kick, — even if you're wearing wellies.

Indeed we *do* have a sick bullock tonight. He didn't come to the trough, runny nose, pneumonia maybe. Phone Robbie who says he'll be out after he's had his surgery and his tea.

I'd forgotten about the tractor battery, — I swear he rose a yard straight up out of his boots.

Friday November 27th

Hard frosty day. The clarts are a lot stiffer. Put the new hoggs onto the turnips.

Robbie comes to visit the pneumonia patient who is improving, and has a check through the other bullocks. They seem to be OK.

At night Gladys and I go to Willie's school to meet the unfortunate staff whose thankless task is to drum a little knowledge into our son. All the teachers are unanimous that it's not easy. At geography he's bad, history below average, maths poor, English diabolical and French pathetic. Shows no aptitude for art, and is only interested in certain aspects of biology (and we can guess what they are!)

Only the PE teacher thinks he's a bright, enthusiastic, talented lad. His mother thinks he's lovely too.

I drop her off at Hilda's (where no doubt they'll be planning to clean out Oxford Street) and go on to the Swan. Charlie has come on his new (insurance) tractor, — he's like a kid with his first bike. Mrs Simpson comes in with a new boyfriend, — she's like a kid on her first date (she always is, maybe that's her appeal).

Saturday November 28th

It's a hard frost again, but not too hard for the cavalry. They pass me by today, Charlie has them for a while, then they disappear over the hill, running north. Only the occasional sound of the horn, and the hounds yapping as they head for some other poor peasant's turnip field.

They only hunt *here* when it's bottomless.

The roads are jammed with supporters, followers, groupies,

— in Volvos and green wellies, carrying Japanese binoculars, English shooting sticks and Scotch whisky.

'They've gorn right hand,' shouts some Barboured expert, and those who know and those who wish they did, head off in the general direction of Arthur Thompson's. Those who don't care, glance at their watches and head for the Swan.

Sunday November 29th   *1st Sunday in Advent*

The bullock is back to normal, eating well, licking himself, stretching, all the right signs. He can go back with his mates.

The turnips are disappearing like snow in a fresh, but I think we should get some hoggs away by the turn of the year, and that will relieve the pressure. They look really fit and well this morning, — it's a dry cold windy day, the fleeces are puffed out, and beginning to colour. The outlying bullocks down by the wood are doing nicely too, — they're a good buy I think. What do they say, — 'well bought is half sold. ... '

Careful, you're getting too cocky, — just let the water run under the bridge.

Monday November 30th   *St Andrew's Day*

I think the tups have completed their annual orgy. They look knackered. I suspect if a yow tried to seduce one of them today she wouldn't have much luck. Anyway what we've got to do now is keep the yowes alive till lambing time. I will be mildly astonished if they all make it.

Jake the postman, who knows everything (and if ever the KGB need a rural spy, this is their man), tells us that the rustlers have been operating again. A farmer over near Jimmy Fletcher woke up yesterday morning to find all his sheep very agitated, and ten hoggs missing. At first he just imagined they'd wandered off somewhere, but there were wheel marks in the field, a bit of sawdust lying, something had caught the gate post. He'd definitely had visitors in the night.

The police were there this morning scratching about for clues and taking details of the stolen goods, — but the stolen goods will probably be chops by now!

Jake has his suspicions, reckons he's pretty sure who the villains are.

Florrie quickly tells him to say nowt, because if he's wrong he could be sued for 'defecation of character' (and nobody would fancy that).

December

Tuesday December 1st

Got the hoggs in off the turnips to draw a few out for tomorrow's mart. It took us all bleedin' morning'!

It was Sweep's fault, — he hasn't had much to do recently, and was 'full of 'imself', — wouldn't stop, went like a train, had the sheep running round in rings with me screaming for him to sit down. (I had a sore throat by nine o'clock.) Of course when we eventually got them to the gate he wouldn't let them out, would he? Sat in the bloody gateway trying to look intelligent. I threw m' stick at 'im, some stones and m' hat (that's never a good idea). Then some of the hoggs bolted up the dyke. Sweep went after them, so the main bunch

stampeded through the gate and away up the road. A maniac in a car started blasting his horn at them, and about half of that group panicked through the hedge in amongst some yowes.

Now we had some on the road, some others in the wrong field being chatted up by a tired tup, and the rest still on the turnips. It took me, Gladys and Charlie (who'd come down to see what all the swearin' was about) and the out-of-form dog until dinner time to get them all together again and into the pens. The tup came too.

Eventually we marked twenty as gradable.

Wednesday December 2nd

Hoggs all graded bar one (too fat) but a moderate trade.

Charlie had about a dozen arthritic old yowes there, and one of his anorexic cows. He got a hellova trade. All the knackered stuff was on fire.

I didn't know at the time, but after he'd dumped his cargo at the mart he left his horse box in the car park and took the Land Rover along to the garage for its MOT. It was only at the end of the day, when he went to yoke up the trailer again to come home, that he found it wasn't there. He couldn't believe it. Had he left it somewhere else and just forgotten? Had he taken it to the garage? Had someone else picked it up by mistake?

No chance, — it was pinched!

Thursday December 3rd

Gladys is packing for London.

In order that Doreen and Willie should not die of malnutrition while she's away the fridge is crammed full, the cupboards are overflowing, and she can't get the lid shut on the deep freeze.

She and Hilda are on the phone every twenty minutes comparing wardrobes. You'd think we were staying with the Queen.

Brought the tups into the croft, feet dressed, de-harnessed, duty done.

They've got a year to get over it.

Friday December 4th

My preparations for the trip to London are simple enough, — bed the cattle with clean straw, crush enough barley meal to last a week and bag it up, — move the nets on the 'break', and generally make it as easy as possible for Gordon to look after the 'empire' while I'm away.

At the Swan all the chat is about thievin'. All year livestock and the occasional piece of equipment have been disappearing regularly. Almost every peasant in the parish has had something pinched ... except me, — perhaps I'm a prime suspect!

Percy reckons it's got to be some highly organised gang from the city. There'll be a Mr. Big running the show, he says, — somebody with transport, and a butcher has to be involved somewhere in the plot. When they're caught, Percy says, they should be hanged, drawn and quartered, and put in prison for life!

Saturday December 5th

The hunt (who met out West somewhere) ended up on the holding of Harry the hairy hill farmer. This was unfortunate, because Harry had recently lost seven hens, two geese, and a duck to the fox, — and together with his sons had set off in search of Reynard and revenge.

Consequently when Knickers and the cavalry galloped over the horizon, there were three stiff foxes hanging in Harry's yard. Not that Harry was in the least bit embarrassed by this visitation. He is a very independent owner-occupier leading a very simple, straightforward rural life, who probably doesn't know anything about credit cards and overdrafts. Nor is he particularly familiar with the art of diplomacy.

The hunt, we are told, left hurriedly, encouraged by a three-gun salute fired (just) over their heads.

It's a freezing cold day and I spend the afternoon indoors. Sweep is with me, lying with his nose on the hearth, sometimes looking up anxiously in case Gladys sends him back to the stable.

Sunday December 6th

In Charlie's car to the train for 5.00 p.m. Gordon is left in charge of all things agricultural. Doreen and Willie are responsible for all things domestic. The keys to our car are in Gladys's handbag, the mileometer checked on the pickup, and instructions given that we want everybody at home and the fires lit when we return. And don't forget to feed Sweep!

British Rail is not at her best on a Sunday. The scheduled three-hour trip to King's Cross via York and Peterborough takes five, via Durham, Darlington, Tow Law, Blackpool, Derby, Wolverhampton and Ipswich (and possibly Brighton for all I know). Certainly long before we arrived at our destination the buffet had run out of anything that was edible or drinkable, and we had run out of polite conversation.

A taxi (I think we may have actually bought it) took us to the hotel in Kensington via Heathrow and Southend, and it was eleven before Charlie and I could finally relax with a few nighcaps in the bar. Gladys was 'dead' by the time I staggered into the room.

Monday December 7th

Charlie and I are up at 6.30 for a walk in the park. Plenty
grazing there, but not a sheep in sight, only joggers with red
tracksuits, red noses and red knees.

After breakfast we ride the Earl's Court earthworm, and
are disgorged two stops later into a sea of farmers, — rusty
faces, big hands, tweed jackets, heading for the show.

Inside, the bars are already doing a good trade. Panorama
farmers with briefcases and townie suits talk earnestly to
salesmen. Kids climb into cabs the size of dining rooms, and
pretend. Everybody wanders slowly up the avenues, pausing
to look at some brightly coloured mechanical monster that
costs more than a house.

I see Roger looking at an even bigger plough than the one
he's got. It'll take an even bigger tractor to carry it.

On some stands there are television sets showing a potato
digger working on a sunny day, or a combine harvesting on
a sunny day, or perhaps just an expert talking on a sunny
day. There are wufflers wuffling nothing silently, a reaper
cuts nothing silently, an elevator elevates nowt, silently, —
all the belts and chains oiled and greased and stretched to
perfection. Men in white overalls see to that. It's a warm,
dry, clean, carpeted, tick-tock world, — where nothing breaks,
nobody kicks and screams. Does anybody buy anything? I
wonder.

Charlie and I buy a couple of pints, and a rep who mistook
incredulity for interest, forces gin 'n' tonic upon us. Didn't
do him any good.

We met Cecil in one of the restaurants. He's ordered a new
tractor, 25 per cent off list, and interest-free credit for three
years. Oh so that's how it's done.

Tuesday December 8th

We took the wives out to dinner (supper) last night at some
very fancy eating house. The Pillicks recommended it to

Gladys, and obviously it was their subtle way of getting revenge for their knackered gnomes.

You could tell it was a mistake as soon as we went in. Not another farmer in the place, all suits and long frocks, waiters buzzin' about like flies on a midden, a menu the size of the *Telegraph*. We did it properly though, went down the card, — soup, fish, meat, pudding, and four bottles of wine.

When the bill came, Charlie asked for 25 per cent off list and interest-free credit for three years, — but they didn't seem to think that was very funny.

Back at the show today we concentrated on the livestock. Cattle waddling about, pampered, perfumed, oblong. All real good 'n's of course, — no bellies, no ringworm, no hook bones. All panting in the heat of this five-star hemmel, and doomed to be judged twice.

Sheep. I didn't know there were so many breeds of sheep. All sizes and shapes, — Blackface, Suffolk, Leicester, Dorset, Clun, and many more. The most surprising thing of all ... not a dead 'n' in sight!

We were supposed to meet Gladys and Hilda back at the hotel, — but we were led astray by some scoundrels from north of the border. It was just going to be a little celebratory nip (one of them had won a prize ticket for a pair of hoggs) but it got a bit out hand. Nip followed nip, pub followed pub, and then we got hungry. After a steak somewhere, we were dragged (honest, Charlie and I protested) down some steps into the bowels of Balham. It took twenty minutes and fifty quid before we could even see what was going on. Actually nothing was going on, it was all coming off.

It cost us about three good cross-Suffolk hoggs before we got out of there!

Wednesday December 9th

We expected the womenfolk to be in bad fettle about last night, — but they were positively glowing this morning. It seems they waited in the bar for ages, gradually becoming a bit silly, and then were chatted up by a pair of rich Warwickshire farmers on the loose.

'Proper gentlemen, beautiful manners, real style,' they said giggling at each other. 'Expensive suits, clean fingernails, knew how to treat a lady, − it was a revelation,' they said. 'Had a lovely night, didn't we Hilda. ... '

Charlie and I said nowt. They were still twittering on about those two Warwickshire weirdos when we pulled out of King's Cross. However, a sandwich and a cup of BR coffee soon quietened them, and by Grantham we were all dozing.

London in small doses might be alright, but not too often.

The trouble with Smithfield week is that the locals see us coming. They can recognise a peasant a mile away, and up go the prices. Somehow we don't blend in with the city scene. Is it the face, the hands, the boots, the 1947 jacket ... or just the general impression of a lost soul?

Thursday December 10th

Back to the real world again. Into wellies and the working gear, − warm and comfy again. We got home just before dark last night, so I had time to check all the stock. No problems.

There's a cheque from the mart in a nice big white envelope, and a lot of nasty little brown envelopes as well (we all know what's in them). The cheque goes straight to the bank, the little brown things go into the bottom left-hand draw of the desk. They can chat among themselves for a while.

Friday December 11th

A friend of Arthur Thompson lost a bullock while he was at Smithfield. Nobody's got it on their farm, − the beast's vanished.

Saturday December 12th

A mild, almost springlike day. You can never depend on this climate. Willie plays rugby as usual, he's been dropped into the second XV this week, and is acutely dischuffed, threatening to ask for a transfer! Where to?

At home we have a visit from the hunt, and four farmers (who should know better) take a short cut across the winter wheat. They'll get a piece of my mind the next time I see them at the mart.

Doreen and Roderick the randy have the weekend off. They wanted to go somewhere for a 'romantic' little holiday, but Gladys put a stop to that. I think she imagines they're less likely to get up to any 'mischief', as long as they stay within a five-mile radius of home (huh).

Willie comes home in triumph, — three tries and a penalty. He's taken himself off the transfer list.

Sunday December 13th

The weather is still remarkable, — many a day in the summer wasn't as good as this, — so we grabbed the chance, and went to do battle on the golf course.

Arthur and I were quite prepared for an expensive humiliation again as usual, but golf, like life, is full of suprises. The surprise today fell upon Rodney. Rodney the six-handicap solicitor, the bore who always hits the ball out of the parish as straight as a bullet, the creep who always chips stone dead, the clever git who always holes the five-foot put. He it was who fell victim to the dreaded socket!

It happened at the third hole. A seven-iron shot flew off at right angles into some bushes. At that stage he wasn't worried, it hadn't happened to him before.

But we ordinary mortals knew that the devil himself was on the course. His provisional ball went the same way. 'Oh bad luck,' said Fletcher. 'Whoops,' said I. 'Oh dear,' said Arthur.

It got no better over the next few holes, and of course the opposition didn't help much either, offering perhaps a little too much advice and sympathy. By the turn he was only using a five wood, and even that without much confidence, — walking along, head down, consumed by unfamiliar fears. And then at the twelfth, he twitched an eighteen-inch putt four feet past the hole, and ran after it!

At the end of the round Rodney was an old man, and we were a quid richer.

Monday December 14th

Gladys is in a huff, I can tell. It's not just the breakdown in communications, — I can stand that, no bother. I don't need a conversation before twelve noon anyway. But it's obvious that something's bothering her. The routine is familiar enough, — dinner plonked heavily on the table, the vacuum switched on while you're listening to the weather forecast, and the wellies placed outside the back door while it's raining.

We might never have got to the core of the problem had it not been for Florrie, — she was putting her coat on to go home when she asked me, 'so what do y' think of the new hair style then?'

'Oh you look very smart,' says I, carefully ignoring the fact that she had curlers in as usual. (They're a fixture, it'll take surgery to get them off.)

'Not me y' fool,' she says, 'your wife. . . . Don't tell me you never noticed. She had it done for London last week. . . . '

Oh.

Tuesday December 15th

A cold, wet day. The frost has gone out of the ground, so the hoggs are up to their bellies in clarts again, and don't look half as good as they did a few days ago. The yowes look lean, — I think we'll have to give them a nibble of trough feed to keep them going.

Wednesday December 16th

Christmas show at the mart. Sent a dozen hoggs, but they weren't good enough for the judging. You have to spend some time tarting them up for that exercise. Ours were straight off the break with a lot of clarts hanging from their bellies, — hardly star material.

Everybody turns up for this sale.

Cecil had a good bullock that he'd obviously been grooming since September. Arthur presented a pair of smart heifers that might've won a ticket, if one of them hadn't been mad-a-bullin' for the last couple of days. Charlie drew out a pen of good-lookin' old yowes, fit and clean, and none of them lame (we've never had yowes like that).

The aristocracy put in an appearance (probably because the dinner was free today). The Ayatollah was unusually affable, dressed in the traditional upper-crust costume of shiny boots, Barbour coat, tweed cap, cavalry trousers at half-mast, and carrying a brand new stick.

Even Ghadaffi the grader was in a good mood. Perhaps he'd been primed with Yuletide spirit. He didn't reject anything, though he did knock three kilos off our hoggs for excessive clart.

I told Gladys how much I admired her new hair-do. She just gave me one of the auld-fashioned looks, and still plonked m' supper down.

I retrieved the wellies before we went to bed.

Thursday December 17th

Feeding the yowes can be a time-consuming job, — there's always at least one lame old bitch who takes an hour and a half to hobble to the troughs. The others are all there, waiting, ready for their nosh (probably been waiting all night) but old twinkle toes is at the far end of the field of course.

There's one of these crippled creatures in both of the yow fields. I could put them into the croft with the tups I suppose, but the rest would just draw lots to decide who should take their place, — and the croft would be full in a week.

Friday December 18th

The brothers Pringle have been arrested for rustling!

Florrie says it's in the papers, — but not until it's confirmed by Jake the postman can we believe it to be gospel.

Jake comes at eleven. They're in custody, he tells us. The police are at the farm now, making lists, and any farmer who's lost a bullock or a baler over the last twelve months is invited to go and have a look.

Charlie finds his horsebox, repainted and with a new number plate, but he'd carved his initials inside somewhere, and they were still there.

Somebody else had already retrieved a chain saw, another found his missing trailer; a long-lost forage harvester turned up. The constabulary found a wagon with no plates or licence, hidden out of sight in a wood, and a Ford Escort covered by a stack of bales.

At the pub the talk was of nothing else. Everybody who had ever lost anything since the war was blaming the Pringles. Gordon remembered a watch that went missing in the winter of '63; Mrs Simpson, a priceless necklace given to her by an admirer shortly after the coronation. Fletcher came up with a set of golf clubs nicked from the boot of his car. Arthur's wife claimed she'd once lost a week's washing off the line. Davy Scott's collie disappeared two years ago. The Pringles got the blame for everything, and farmers who stole from fellow farmers could expect no mercy.

Percy's plan to get them out of gaol and string them up

immediately, attracted considerable support ... but Elsie called 'last orders', and anyway it was a right moderate night outside.

Saturday December 19th

Gladys has discovered she needs more Christmas stuff, and there's only four more shopping days left. She's speaking again, she's even smiling again, and I have no reason to suppose this is because she wants more money again. Of course not. It doesn't really matter, life's a lot easier when she's in a good fettle, and what's more the wellies are back indoors again.

And Willie's back in the first team. He had a good game. They won, but he got knocked about a bit, and stayed in at night, — resting. Tracey came to 'rest' with him in the kitchen. I went through for a cup of tea after Match of the Day, and I don't think they'd been resting properly. They both seemed out of breath t' me.

Sunday December 20th

I think I've got flu, and there's no doubt in my mind where I got the bug ... London! All those coughin', wheezin', splutterin' townies who spend their lives in hot tubes, hot hotels and hot offices, have passed on their nasty urban bacteria to me!

Gladys has little sympathy, threatens to phone the kennels, and reminds me of the other (rare) occasions when I've been under the weather. She even resurrects the great measles fiasco of '75. It wasn't funny really, but I was sure I had a bad attack at the time, felt awful, couldn't stop scratching.

It turned out to be hayseeds inside m' string vest. She never lets me forget.

Anyway I bravely struggled round and fed the 'zoo' before

going back to bed. I didn't get up for the roast beef ... so I must be ill.

Monday December 21st

Fed all the stock with a sore throat. Couldn't even scream at the dog. I didn't wait for the lame yowes either, — I'm in a worse state than they are. Then back to bed. Willie fed the cattle at night.

Willie is on holiday again, and came home with a report that confirms our growing suspicions that he has little chance of being Chancellor of the Exchequer some day.

Gladys is still shopping, this time for wrapping paper. We have seventy-two Christmas cards so far, eight from people we forgot about, and four from folk we've never heard of.

Doreen came up to the bedroom to pay her last respects after supper. I told her to be brave, and look after her mother.

Later her mother forced me to drink a terrible concoction of hot whisky, lemon and aspirins. Soon after that I drifted into oblivion, convinced she'd poisoned me.

Tuesday December 22nd

Quite surprised to wake up at eight o'clock, feeling much better. I put my dramatic recovery down to a stiff upper lip, and a stout peasant constitution.

Investigations continue into the Great Rustling Scandal. Nobody round here cares about rampant fraud in the city, or insider dealing on Wall Street, — that's par for the course, — but robbing peasants is big news!

Most of the stolen machinery has been recognised and reclaimed, but the missing livestock is a more difficult problem. Jake tells us that a few cattle have been identified, those with tatooed lugs for instance, — but most of them have just vanished. The Pringles have been selling the stores at far-away marts, and the fitter stuff must've been slaughtered straight away.

As Florrie says, 'the incinerating evidence is gone. . . . '

Back on the ranch it's not a good day.

The outliers are in the wood. All bar one find their way out again as soon as I make a noise like a bale of hay, but this lone bewildered beast just gallops along the fence (past the gap where he got in) with me panting and spittin' after him. Sweep is no use in such circumstances. He's easily confused, eventually becomes depressed at my raving obscenities, and goes home long before I get the bullock back into the field.

There's a hogg away by itself on the turnips. I expect it's preparing to die for Christmas.

One lot of yowes have developed a taste for winter barley, — even the acutely lame have managed to stagger through the dyke. Predictably, they can't stagger back again and have to be carried.

The afternoon is spent repairing fences, and I end the day with a black throbbing thumbnail.

The film on telly tonight is 'The Sound of bleedin' Music', — again!

Wednesday December 23rd

A cold frosty day, and McCaskill is suggesting the possibility of a white Christmas. He's not actually forecasting it, y' understand, — he's not *that* confident. He says it'll be cold (reasonable) and there *could* be scattered showers, — which may fall as snow on high ground.

Gordon is much more positive. It will definitely snow he says, — the conviction is based on the behaviour of a mole in his garden. — It's digging south, — it's a sure sign he says.

That hogg has gone back into the pack. I can't find it. Perhaps it'll live after all.

Thursday December 24th

Christmas Eve. There's a tree in the corner of the living room, covered in tinsel, crackers, stars and fairies. Grandad's photo, the picture of the highland cattle drinking in the Tweed and the mirror over the fireplace are all obscured by holly. A hundred cards blow off the sideboard every time somebody opens the door. The bowls are full of friut, nuts, dates and toffees, — but Gladys gets upset if we touch anything before the big day.

Took the pickup into Northern Grain for yow nuts, and spent half an hour in Woolworths doing my own Christmas shopping. It took another half-hour to pay for them at the cash desk, before I could get out again. (The world's gone mad.)

The livestock seem oblivious of all this nonsense, all *they* want is food and water as usual.

I've no idea what's wrong with that hogg, but something's not right, so I inject the beast with a bit of everything I can find left over from the lambing.

Friday December 25th   Christmas Day

Happy Christmas, ho, ho, ho.

The tractor with the link box on won't start. The other one has a flat front tyre. Fortunately it'll pump up, and hold long enough to do the jump lead trick, — but the yowes and outliers are blarin' impatiently by the time I get to them, half an hour late.

Over breakfast we exchange presents. The perfume I give Gladys is the wrong kind, but Doreen seems eager to have it, instead of the bracelet I bought for her. Gladys says the bracelet's 'quite nice'. The tool kit for Willie goes down fairly well, — he says it's much better than the one I gave him last year, which he subsequently swopped at school for a pair of rugby boots.

And a tin of Chum for Sweep, — who's the only truly grateful recipient.

Personally I was overwhelmed to receive two pairs of wellie socks, a ballpoint pen that leaks and a bottle of fizzy white wine, Chateau bottled in Yorkshire, and priced attractively (for Willie) at 95p a litre.

We all ate far too much, and eventually collapsed to watch 'Bridge on the River Kwai' (again).

There was no snow.

Saturday December 26th   *Boxing Day*

That hogg on the turnips has lambed!

It's a unique creature the size and shape of a retarded rabbit, — but it's alive and full, and mother seems delighted.

Roderick comes for tea bearing a gift for Doreen. Doreen is reluctant to reveal the contents of the box, but Willie later confirms it to be a mere morsel of lacey lingerie. What Florrie would call 'erratic underwear'.

Roderick is no fool. He brings Willie a present too, — a multicoloured T-shirt.

Roderick is extremely cunning, — he brings Gladys her favourite smelly stuff.

Roderick is a very canny lad, — he brings me a bottle of Scotch. He could go far, that fella.

Sunday December 27th

Another hogg has lambed twins!

According to my diary, that tup was over the fence for no more that five minutes in August. He must have performed on the run.

Gladys persuaded us all to go to church, and this being the season of goodwill we didn't argue too much. Everybody was there, singing, 'Away in a Manger', and 'Good King Wenceslas' as hard as they could. Well, at least the women and children did, — the organist is always three octaves too high for the blokes.

Monday December 28th   *Bank Holiday*

Brought the two hoggs and their unexpected offspring into the croft.

At first I tried to simply walk them out of the gate, but of course all the rest went through quick as a flash, and I was left standing on the turnips with two hoggs, three lambs and Sweep.

Moved on to the next plan. Catch the lambs and try to get the mothers to follow. One of them did (with me bleating) but only so far. When I dropped the lambs to open the gate, mother fled.

Enter Willie, attracted by colourful language, and on to

plan three. Put the lambs into the link box, and then catch the two hoggs.

Sweep gathered the whole lot into the corner, and Willie and I made a rush for one each. He caught his, but I missed the tackle, got trampled half t' death, and covered in clarts. Put Willie's victim into the link box (legs tied) and did the grabbing in the corner trick all over again. We were supposed to be going for the same hogg, but Willie got one and I got another. He got the right one.

He'll likely tell his mother about it.

Tuesday December 29th

Cold grey day, — it feels like it might snow before dark.

Got up at 7.30. Gladys dreaming like a dog who imagines he's chasin' a rabbit, — little squeaky yaps, and the back leg jumpin'. God knows what she's up to. . . .

Thankfully no more lambs, and everything comes to the troughs, eventually.

The bullocks are doing well. They look especially good when the bedding's clean. We'll start selling a few when this turkey-eating festival is over. It'll be a week yet before people get back to work and a normal diet.

A few flakes of sloppy snow falling at bedtime.

Wednesday December 30th

There's been a gentle three inches of the white stuff over night, and the sky looks full. The old tom cat is waiting at the back door, and bolts in as soon as it's opened, — his footprints the only marks on the flat blanket that covers the yard.

Spuggies and starlings are chattering in the hemmel, sitting on the beams, some looking for scraps in the trough, or scratching among the bedding. The first sound I make (a rattling pail, a creaking gate, a cough) is the signal for the cattle to stretch and shout, — and the day to begin.

Some peasants will have been up and working for a couple of hours already, like the ones who milk cows fourteen times a week. Some might have been up all night farrowing a sow, calving a cow. Others will lie abed for a while yet, while the snow flutters down onto their quiet arable acres.

Charlie will be setting off to feed his ravenous suckler cows (I can hear them). Arthur will be corning his yowes (he seldom turns the troughs, so they'll be full of snow this morning). Jake the postman will be struggling up the road to Fletcher's place, knowing full well that the two technicolour circulars he has to deliver will go straight into the fire. But he'll get a cup of tea for his bother.

Robbie the vet might just be coming home from a caesarian. The Council men will be gritting the A1 on double time. Gladys is frying the bacon, I can smell it.

Thursday December 31st

The Old Year's last fling.

Nowt special about it really, — it's just a moderate winter's day, and dark by tea time. The stock still have to be fed, and (y' never know) some auld yow might consider it a suitable occasion to have *her* last fling.

As it happens everything is present, and more or less correct.

I spend the afternoon crushing barley, moving hay, loading the link box for tomorrow. I may not be at my best in the morning.

By early evening the telly is already filled by kilted Scotsmen singing silly songs about purple heather, and bonnie wee lasses roaming about the heelands.

We're getting ready to go out (and who can blame us).

Doreen and Roderick are off to a sophisticated soirée in suburbia. She'll freeze to death in that outfit on a night like this.

Willie and Tracey are bound for a teenage orgy at the Rugby Club, with strict instructions from Gladys to behave like responsible adults.

The responsible adults go to the Swan, and then up to Charlie and Hilda's where they eat too much, drink too much, and talk a lot of rubbish.

At the stroke of midnight Gladys finds me in the kitchen. It's tomorrow already. It's next year already.

'Happy New Year, pet,' she says.